HUBBLE

NASA Images of Planets, Stars, Galaxies, Nebulae, Black Holes, Dark Matter & more

Amherst Media, Inc.   Buffalo, NY

AUTHOR A BOOK WITH AMHERST MEDIA

Are you an accomplished photographer with devoted fans? Consider authoring a book with us and share your quality images and wisdom with your fans. It's a great way to build your business and brand through a high-quality, full-color printed book sold worldwide. Our experienced team makes it easy and rewarding for each book sold—no cost to you. E-mail **submissions@amherstmedia.com** today

Published by:
Amherst Media, Inc., P.O. Box 538, Buffalo, N.Y. 14213
www.AmherstMedia.com

Publisher: Craig Alesse
Senior Editor/Production Manager: Michelle Perkins
Editors: Barbara A. Lynch-Johnt, Beth Alesse
Acquisitions Editor: Harvey Goldstein
Associate Publisher: Katie Kiss
Editorial Assistance from: Ray Bakos, Carey A. Miller, Rebecca Rudell, Jen Sexton-Riley
Business Manager: Sarah Loder
Marketing Associate: Tonya Flickinger

ISBN-13: 978-1-68203-300-5
Library of Congress Control Number: 2017958111
Printed in The United States of America.
10 9 8 7 6 5 4 3 2 1

www.facebook.com/AmherstMediaInc
www.youtube.com/AmherstMedia
www.twitter.com/AmherstMedia

Contents

An Introduction to Hubble Images 5
The Hubble Telescope 6

Hubble Service Missions 8

Seeing Space
Through the Hubble Telescope 10

Hubble Sees the Moon 12
What Wavelengths Can Hubble See? 12
Hubble Telescope Instruments 13
Stellar Magnitudes 14
Types of Magnitude 14
Light and Dark 15
The Colors of Space 16
Seeing into the Distance 18
True and False Colors Reveal 19
Pixels and Resolution 19

Our Solar System 20

No Hubble Images of Mercury 20
Venus Clouds 21
Earth 22
Mars and Polar Ice 24
Mars and Tiny Phobos 27
Jupiter's Moons and Their Shadows 28
The Great Red Spot 29
Ultraviolet Images of Comet Impacts 29
Jupiter and the Colliding Comet 30
Jupiter and Comet Shoemaker-Levy 9 30
Europa, Jupiter's Moon 31
Saturn's Changing Seasons 32
Saturn's Rings 34
Uranus 36
Bright Clouds on Uranus .37
Aurorae on Uranus 37
Neptune's Weather 38
Neptune's Day 39
Pluto and Its Moons: Charon, Nix, and Hydra 40
Pluto's Tumbling Moons 41

Our Solar System's Comets 42
Sungrazing Comet 43

Stars, Supernovas
& Planetary Nebula 44

N6946-BH1 Became a Black Hole 44
PIA03519: Cassiopeia A 45
Signs of Star Formation 46
Ring nebula (M57) 47
Stars and Globules 48
Protoplanetary Nebulas 49
One of the Hottest White Dwarfs 50
Cat's Eye Nebula: Beautiful Dust Shells 51
Helix Nebula: PIA03678 .52
Eta Carinae: Twin Stars 53
Death of a star: Calabash Nebula 54
The Crab Pulsar and Supernova Remnants 55

Black Holes and Quasars 56

Difficult to See 57
Fueling a Black Hole 57
Super Massive Black Hole 58
Obscured, Active Galactic Nucleus 59

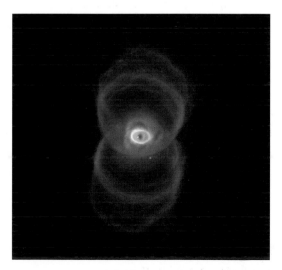

Dark Matter and Dark Energy **60**

Dark Matter in Galaxy clusters 61
Mysterious Dark Energy. .62
Measuring Dark Energy. .63

Our Galaxy, The Milky Way **64**

The Arches Star Cluster 65
Hen 2-437 65
Our Neighbor, Super Star Cluster Westerlund 1 . .66
A Filament of SN 1006. 67
Sagittarius A: Our Own Galaxy's Black Hole 68
A Closeup 69
Stars of the Milky Way 69
The Milky Way Using Three Telescopes 70
Milky Way Stars: Birth and Death. 72
Herbig-Haro 24, HH 24 .73
Pillars of Creation. 74
Tower of Gas 75
A Stair Step Image 75
Infrared Shows Brilliant Stars 76
Old Stars in the Milky Way 77
Horsehead Nebula 78
The Orion Nebula. 80
Carina Nebula 81

Galaxies **82**

A Twisted Disk 82
Hubble Deep Field 83
Ultra-Deep Field 84
Ultra-Deep Field: Refined Images 85
Whirlpool Galaxy 86
A Backward Moving Spiral Galaxy 87
Small Magellanic Cloud. .88
Starburst Galaxy 89
Cartwheel Galaxy. 90
Star Forming Activity Spurred 91

Bubbles Rising from the Core 92
Hubble-V in NGC 6822 .93
Hubble-X Star Formation Burst 94
Messier 104. 95
A Dwarf Galaxy Puzzle: NGC 1569. 96
NGC 1569 Is Moving Away. 97
Galaxy nGc 1512. 98
Wavelengths Recorded. .99
Center Ring. 99
Merging Galaxies 100
Galaxy Messier 101 102
Merging Galaxies II Zw 096 103
A Galaxy Pair: NGC 454 104
Graceful Impact: Hercules Galaxy Cluster. 105
Collision Between Disk Galaxies 106
Edge View of a Spiral Galaxy. 108
Cosmic Dust Bunnies 109

Galaxy Clusters: Gravitational Lenses 110

Hubble Deep Fields 111
Cluster Abell 1689 111
Wide-Field Image 112
Ground-Based Abell S1063 113
Parallel Field of Abell S1063 113
A Massive cluster 114
Frontier Fields Program .115
A Happy Face Cluster. 115
Galaxy Cluster Abell 1689. 116
Looking into the Past 117
Cluster SDSS J1110+6459 118
Dark Matter Map Galaxy Cluster 119
Abell 2744 Cluster Lens 120
Intracluster Space Light. 121
Multiple Galaxy clusters. 122
Mapping Mass 123

The Future of Space Telescopes. 125
Index **126**

Hubble images from space are amazing. The images in this book provide enjoyment, inspiration, and an egress into NASA Hubble Space Telescope's exploration of outer space. The image comments are short, succinct, and in non-technical terms, providing enough context to get "space novices" started. More in-depth and detailed information can be found on the internet.

The NASA Hubble Space Telescope images are simply beautiful. Each one is different and describes worlds in our solar system and far away places. They span across unimaginable distances—most are light-years away—to planets, stars, supernovae, black holes, galaxies, and nearly into the birth of our universe. They display for us the distant, the invisible, the incredibly hot and cold, the long ago, and the very beginning. Hubble images peak our curiosity and wet our appetite for solar and galactic wanderlust.

These scientifically recorded, data-based images are artfully created by NASA and their many associates. NASA and its communities share Hubble's images, and the data collected, with the world. The images in this book are only a small representation of what has been collected. In fact, there are huge archives of data waiting to be reviewed for scientific studies and image generation. Hubble images and their corresponding data have answered many scientific questions. And as it often happens when answering science questions, more important questions arise—about space, physics, life, and the universe. Many of Hubble's classic images have been updated (and will continue to be) with refined imaging techniques or new data from other telescopes and instruments. Many scientists/artists continue to work as data is still being collected and the data's usefulness will continue to add to our enjoyment and knowledge of the universe.

As the curator/editor, I have tried to give image credit as requested by the websites where each image was acquired. Similar images were available from different websites with variations in credits provided. If I have left anyone out, please contact me and corrections will be made in future editions. Also, if you would like to explain your imaging process, feel free to reach out. For those whose images were not included, I would like to hear from you, too.

More fantastic images and knowledge are yet to come from the NASA Hubble Space Telescope data and other future space telescopes. We are on the cusp of significant discoveries, and the world witnesses it through earthbound and space telescopes' visual platforms. Enjoy this collection of Hubble's astounding images.

Beth Alesse, editor/curator
BAlesse@HubbleImagesfromSpace.com

The Hubble Telescope

In 1609, Italian scientist Galileo Galilei used the newly invented telescope to view the night sky. His observations conclusively changed humankind's grasp of the universe. Nearly four centuries later, the NASA Hubble Telescope, named for astronomer Edwin Hubble, is also changing our fundamental understanding of the universe.

Orbiting 340 miles above the surface of our planet, Hubble can see past the distorting effects of Earth's atmosphere. This effect blurs starlight and blocks some wavelengths of light from ever reaching the ground-based telescopes. Out of the Earth's atmosphere, the Hubble observations become more consistent, detailed, and sensitive. The telescope is in its third decade, having made over a million observations, and has produced over 14,000 scientific publications. Besides becoming a huge worldwide cultural icon, Hubble's scientific discoveries are momentous.

NASA's Hubble Space Telescope was launched aboard the Space Shuttle Discovery in 1990 in a partnership between the United States space program (NASA stands for National Aeronautics and Space Administration) and the European Space Agency. Hubble received hardware and software updates during several service missions, and it continues to function as one of the most significant scientific instruments ever made.

Solar panels provide the power for its operation. It has communication antennas to send its data to scientists on Earth. There are two mirrors, primary and secondary. The aperture door can close if the inside needs to be protected. The fine guidance sensors help to orient the telescope to its target and hold it in place while exposures are made. This is a simple list of the major parts of a complex machine that takes advanced software and numerous scientists with technical skill to maintain and operate.

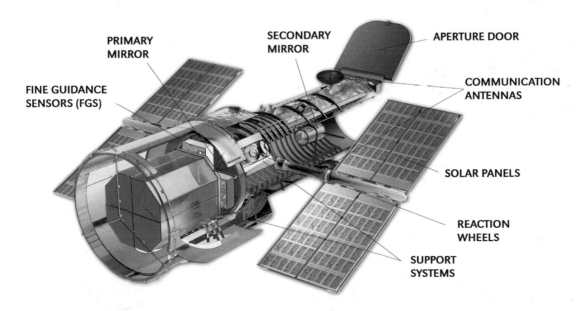

PRIMARY MIRROR

SECONDARY MIRROR

APERTURE DOOR

FINE GUIDANCE SENSORS (FGS)

COMMUNICATION ANTENNAS

SOLAR PANELS

REACTION WHEELS

SUPPORT SYSTEMS

Hubble
Service Missions

The NASA Hubble Space Telescope was designed to be serviced by astronauts in space. Five subsequent Space Shuttle missions replaced, repaired, and upgraded systems on the telescope, including all five of the main instruments by 2009. The Hubble operated with a defective mirror for the first three years until it could be optically corrected. Nonetheless, the telescope carried out productive operations of less difficult targets where astronomers could compensate for the mirror's defects. The fifth mission was nearly cancelled following the Columbia Shuttle disaster in 2003.

The telescope is currently operating in 2017 with plans for use at least until the James Webb Space Telescope (JWST) is launched and functional in 2019. It could possibly last until 2030 and even into 2040. No doubt as long as it remains operational, there is plenty of work for both Hubble and its telescope successor, JWST.

These images *(right)* show the same target galaxy, M100, before *(top)* and after corrective optics had been installed correcting the aberration. As of 2009, all the original observation instruments that have been replaced are functional except for the High Resolution Channel of the Advanced Camera for Surveys (ACS).

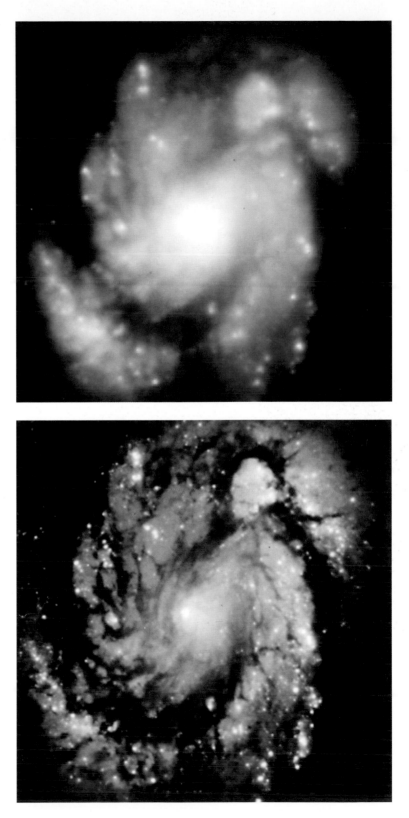

Seeing Space Through the Hubble Telescope

Our human eyes have provided most of what we know about the world and universe. The invention of the telescope increased the distance we can see into space. However, even with the most powerful telescopes, Earth's atmosphere distorts and obscures what we see. A telescope can magnify, but the light waves are diffused and bent by our planet's atmosphere. The Hubble Telescope orbits the Earth above the atmosphere producing images of outer space that are tremendously clearer than its terrestrial counterparts.

Hubble Sees the Moon

These two high-resolution ultraviolet and visible light images *(below)* show the Apollo 17 site in the Taurus-Littrow Valley last visited in 1972. The red Xs in the Hubble images *(below)* show the Apollo 17 landing position. The descent stages that were left behind pictured in the upper right corner, are the size of a small truck and can't be seen by Hubble. The smallest sized object that Hubble can see at this distance is around 60–75 yards.

Hubble can see things that the human eye cannot see. The color composite image of the Aristarchus impact crater *(facing page)* uses ultraviolet-to-visible light information to accent differences in surface materials that may help indicate possible ilmenite-bearing, volcanic glasses, and other materials. It is useful for materials to be identified for future missions to the Moon.

Image credit: NASA/ESA/HST Moon Team

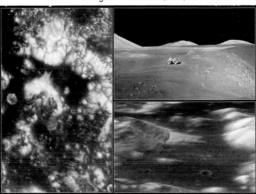

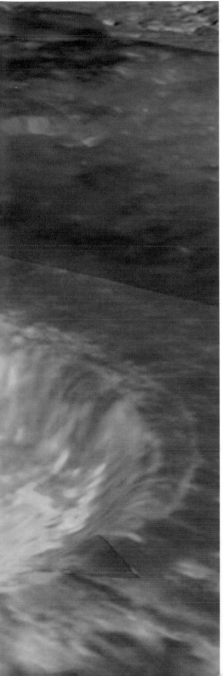

Image credit: NASA/ESA/HST Moon Team

What Wavelengths Can Hubble See?

After NASA's completed service missions, Hubble can see the cosmos using six instruments. Filters are used to record and fine-tune specific ranges of light. The Advanced Camera for Surveys (ACS) can see ultraviolet radiation, visible, and near-infrared. The Cosmic Origins Spectrograph (COS) breaks ultraviolet radiation into smaller parts allowing scientists to study galaxy evolution, for ex-ample, how planets form. The Fine Guidance Sensor (FGS) helps target and lock onto guide stars, measure their relative brightness, and orient the view of the objects being imaged. It helps keep Hubble pointed in the right direction. These sensors can also make precise measurements. The Near Infrared Camera and Multi-Object Spectrometer (NICMOS) is sensitive to infrared light. It can see through the gas and dust of deep space that many objects are obscured by. The Space Telescope Imaging Spectrograph (STIS) was designed to separate and record light into its components, including visible, ultraviolet, and infrared. Unfortunately, it is no longer working. However, the Wide Field Camera 3 (WFC3) can be used to see near and far-infrared, and ultraviolet. So it replaces much of NICMOS and STIS's functions.

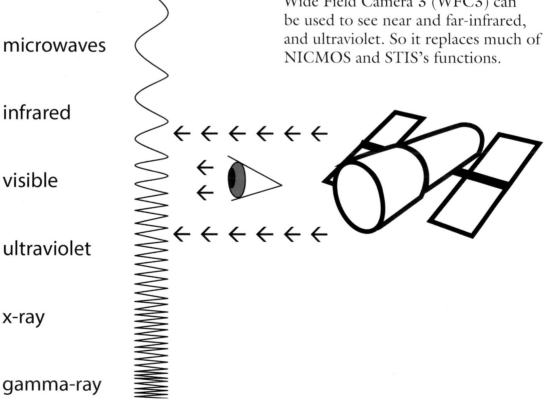

radio waves

microwaves

infrared

visible

ultraviolet

x-ray

gamma-ray

Hubble Telescope Instruments

NASA's Hubble Telescope currently has six instruments:

- Advanced Camera for Surveys (ACS)

- Cosmic Origins Spectrograph (COS)

- Fine Guidance Sensor (FGS)

- Near Infrared Camera and Multi-Object Spectrometer (NICMOS; hibernating since 2008)

- Space Telescope Imaging Spectrograph (STIS; non-operative)

- Wide Field Camera 3 (WFC3)

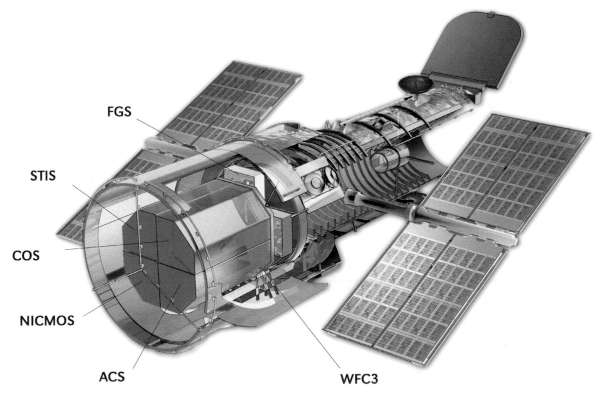

FGS

STIS

COS

NICMOS

ACS

WFC3

Stellar Magnitudes

The ancient Greeks recorded the brightness of stars. They determined the greater the magnitude, the fainter the star—from first to sixth magnitudes. In 1610, Galileo pointing his telescopes into the night sky, discovered that there were stars much fainter than the sixth-magnitude stars, and a seventh magnitude was added. As telescopes got stronger, astronomers added higher and higher magnitudes. Today, the Hubble Space Telescope can see faint objects to the 31st magnitude. We have also recorded objects brighter than first magnitude. The solution to assigning a number to these brighter objects was to add a 0 magnitude, and for even brighter objects negative numbers are used. The Sun (*top left*, taken by Solar Dynamics Observatory), for example, has a −26.7 magnitude.

Types of Magnitude

There are different kinds of magnitude. The *apparent* magnitude is the brightness of an object as it looks viewing it on Earth from the night sky. *Absolute* magnitude is the brightness of an object as if all objects were placed at the same distance from Earth. *Absolute bolometric magnitude* is an object's luminosity over all wavelengths, not only the visible wavelengths. Many of the Hubble images are a combination of visible, infrared, and/or other wavelengths.

Galaxy IC 342 (*bottom left*) is so bright that it can be seen on Earth with binoculars. The bright blue area in this Hubble image are hotter star-forming regions.

Light and Dark

Hubble images often include many types of objects with a range of lumi-

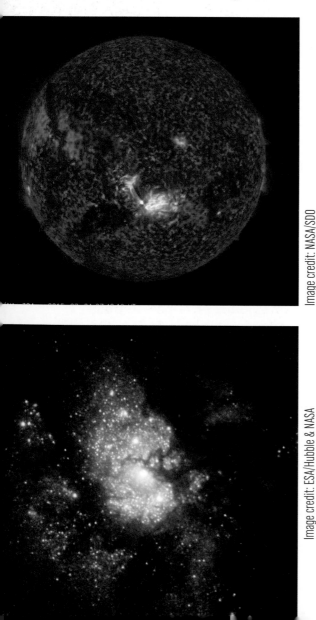

Image credit: NASA/SDO

Image credit: ESA/Hubble & NASA

Image credit: ESA/Hubble & NASA

nosity, details, and brightness. The data recording different wavelengths from more than one instrument are combined to make a final image. So, an informed and nuanced balance of the light and dark values is needed to show what is happening within the image.

In this Hubble image for example *(above)*, in the upper middle is a small young star, SSTC2D J033038.2+303212, with material around it that is really a disk or saucer shape on edge. Below it is a reflection nebula, [B77] 63. Within the nebula are two stars, LkHA 326 LZK 18, which light up the gas inside the nebula. Then, in front of nebula [B77] 63 is a dark nebula called Dobashi 4173. It is called a dark nebula because its materials are dense obscuring most of what is behind it. The stars that are on top of this area are actually in front of it and not part of it.

The Colors of Space

Hubble images often begin as three separate black & white images of an object taken at different wavelengths. When they are put together, each is assigned a color. This composite image is balanced to communicate the scientist's understanding of the object being studied.

Hubble images are often combined with images taken by other instruments as each records different qualities of the object. This image of the Crab Nebula uses data from four other instruments combined with a visible light spectrum Hubble image.

Hubble Space Telescope:
Visible in green

The VLA:
Radio in red

Spitzer Space Telescope:
Infrared in yellow

XMM-Newton:
Ultraviolet in blue

Chandra X-ray Observatory:
X-ray in purple

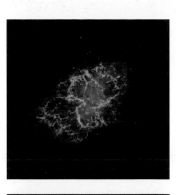

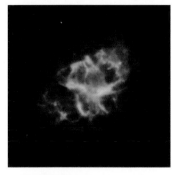

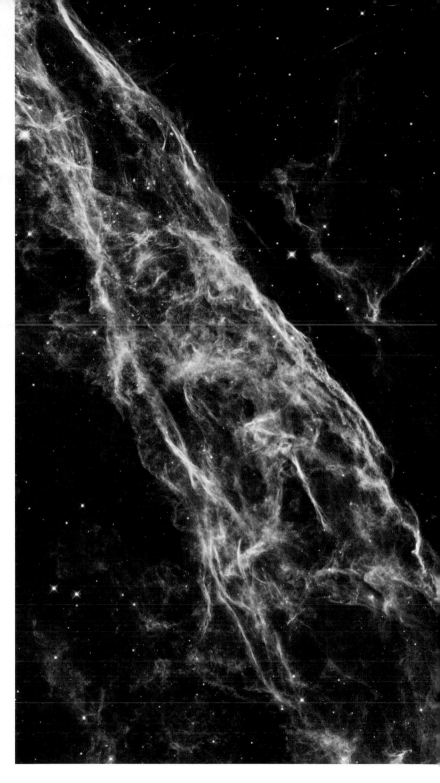

Image credit: NASA, ESA, G. Dubner (IAFE, CONICET-
University of Buenos Aires) et al.; A. Loll et al.; T.
Temim et al.; F. Seward et al.; VLA/NRAO/AUI/NSF;
Chandra/CXC; Spitzer/JPL- Caltech; XMM-Newton/
ESA; and Hubble/STScI

Image credit: NASA, ESA, J. Parker (Southwest Research Institute), P. Thomas (Cornell University), L. McFadden (University of Maryland, College Park), and M. Mutchler and Z. Levay (STScI)

Ceres lies between the orbits of Mars and Jupiter.

Image credit: NASA, ESA, H. Weaver (JHUAPL), A. Stern (SwRI), and the HST Pluto Companion Search Team

Pluto and its moon Charon are 0.000624 light-years away and lie within the Solar System.

Seeing into the Distance

The more distant an object is the fainter its emitted light will be. The naked eye, earthbound telescopes, and even the Hubble, until galactic lensing, cannot see the most distant objects in the universe. Even closer objects in our own solar system become, for the Hubble, less clear the farther they are from Earth. This is true because these objects do not emit their own light but reflect the Sun's light. Surfaces on farther objects in the solar system, Pluto for example, are less detailed than close objects such as Saturn's moons.

The same is true for making images of more distant objects such as stars, galaxies, and galaxy clusters—although these objects do emit their own light. As our instruments and techniques become more refined, our data collection is more successful and the farther they can see. Pictures of far-away objects in the universe add significantly to our understanding of it and to our understanding of our closer world.

It is helpful to know that a light-year is a unit of astronomical distance equivalent to the distance that light travels in a single year. So, Pluto and its moon Charon (*middle left*) may look like stars, but they are only a tiny fraction of a light-year away within our solar system. On the other hand, the image of galaxy NGC 2768 (*facing page, right*) may look smaller than Pluto and Charon, but it is 65 million light-years from Earth and probably contains more than a million stars.

Image credit: NASA, ESA, and
J.-Y. Li (Planetary Science Institute)

Comet 252P/LINEAR is a near-Earth object discovered in
2010. It has an elliptical orbit that can place it, at times,
very close to the Sun.

Image credit: NASA/ESA/Hubble

Galaxy NGC 2768 is an elliptical galaxy 65 million
light-years from Earth. At its center is a super massive
black hole.

True and False Colors Reveal

Instruments on board the Hubble are sensitive to a specific ranges of electromagnetic energy. Each image starts as a singular monochromatic one. As it is combined with the data from other instruments, the colors begin to look more *real*. Some of these images present true colors. However, some of the colors in Hubble images, often combined with data from other instruments, use data that the human eyes cannot see, such as infrared or ultraviolet. Colors are assigned to specific bands in the electromagnetic spectrum resulting in false-color images. It allows us to see what would otherwise be invisible to our eyes. What exists but is obscured becomes tangible.

Pixels and Resolution

Many objects of observation are a very tiny part of the image captured. When zooming for a closer look, the resolution begins to break down or become pixelated. The Hubble's hardware and software upgrades help increase the image quality. As a result, when viewing images that are low resolution, for example, the images above, expect them to be pre-upgraded Hubble imagse or images of objects that are very, very far away and has been extensively enlarged. These pixelated images tell us about the objects of our observations and are nonetheless beautiful and scientifically groundbreaking.

Our Solar System

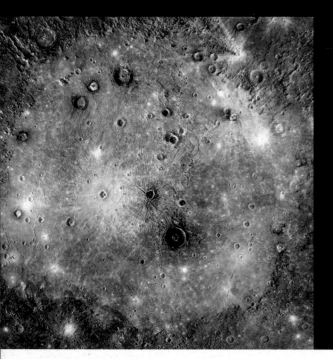

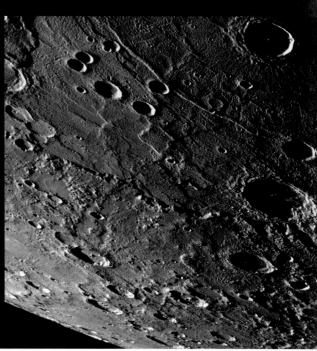

No Hubble Images of Mercury

The planet Mercury is too close to the Sun for Hubble's cameras to take a picture. So these Mercury images *(this page and facing page bottom)* come from probes that visited Mercury. They use similar color imaging techniques as are used with Hubble images produced of farther objects.

Venus Clouds

Cloud tops on Venus were viewed by Hubble. This purple image is a NASA Hubble Space Telescope ultraviolet-light exposure of the planet, taken January 1995, when Venus was at a distance of 70.6 million miles or 113.6 million kilometers from Earth. Notice the low resolution indicated by the visible pixels. This image was taken before all the upgrades had been completed on the Hubble Telescope. Consequently, the image is pixelated and the curve line of the planet appears jagged.

Image credit (facing page and bottom): NASA/Johns Hopkins University Applied Physics Laboratory/Carnegie Institution of Washington

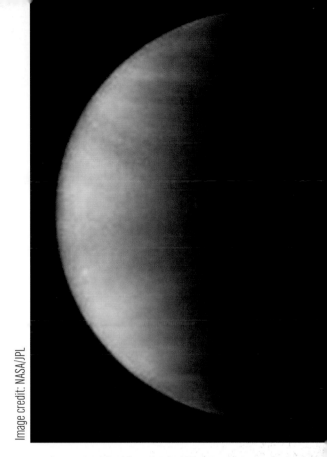

Image credit: NASA/JPL

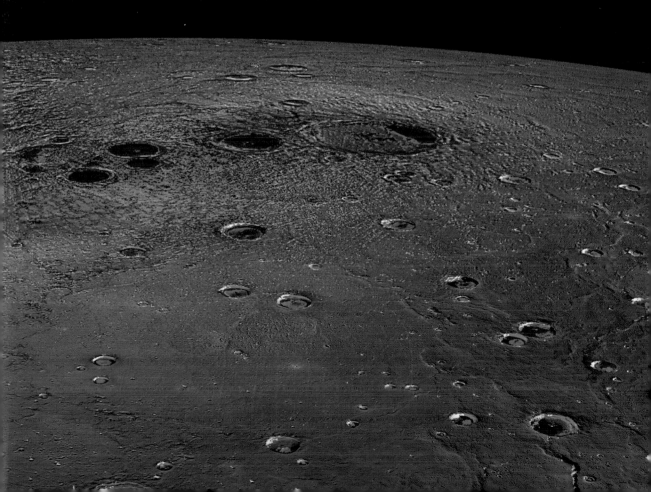

Earth

Hubble is only a little over 300 miles above the Earth's surface and more than seven kilometers per second. Therefore, an exposure would be a blurred one. This low resolution image has motion blur *(below)* that is difficult to eliminate even with tracking because of how close Hubble is to the Earth. Hubble was designed to record objects that are much farther away. Surface tracking of near objects is not its purpose.

There are about 400 geostationary satellites orbiting the Earth at an altitude of approximately 22,300 miles that are more suitable for Earth imaging. They are used for communications, imaging, weather data, and more. This frees the NASA Hubble Telescope to direct its attention to objects deeper in space.

Image credit: Mark Clampin / NASA

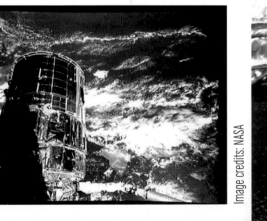

Image credits: NASA

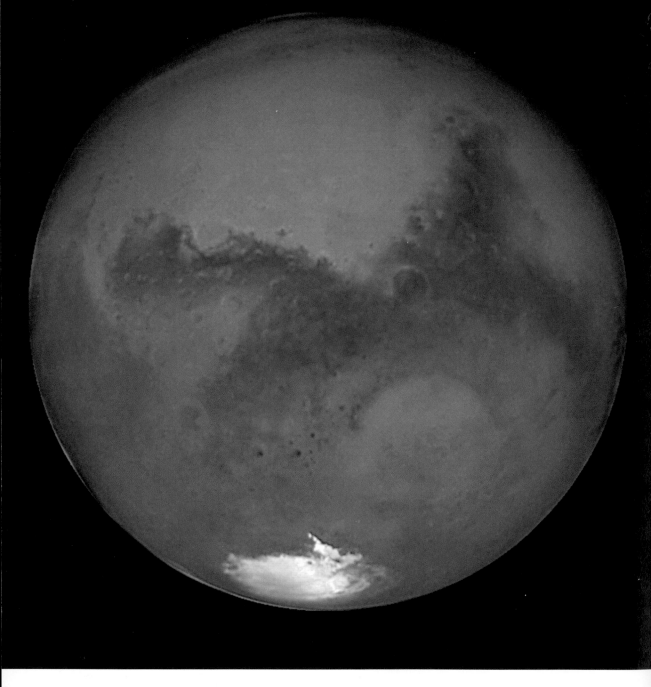

Mars and Polar Ice

The image above shows some of Mars' dark, circular impact craters.

Water ice is on both of Mars' poles. A thin layer of carbon dioxide ice forms during the poles' winter season. Scientists also look for water ice as well. Water is necessary if we are going to create habitats for living on

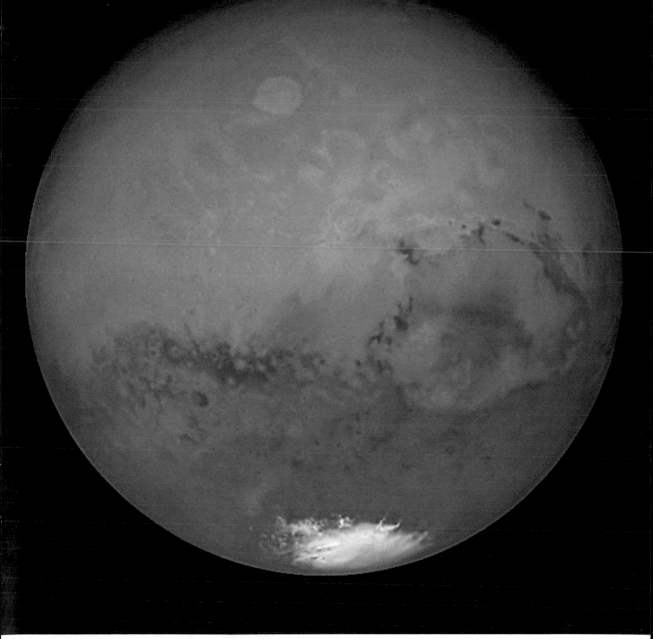

Image credit: NASA, J. Bell (Cornell U.) and M. Wolff (SSI)

Mars. Water is too heavy to transport to another planet. Whatever can be found out about Mars through imaging—its climate, weather, and resources—is an important step in planning such a mission.

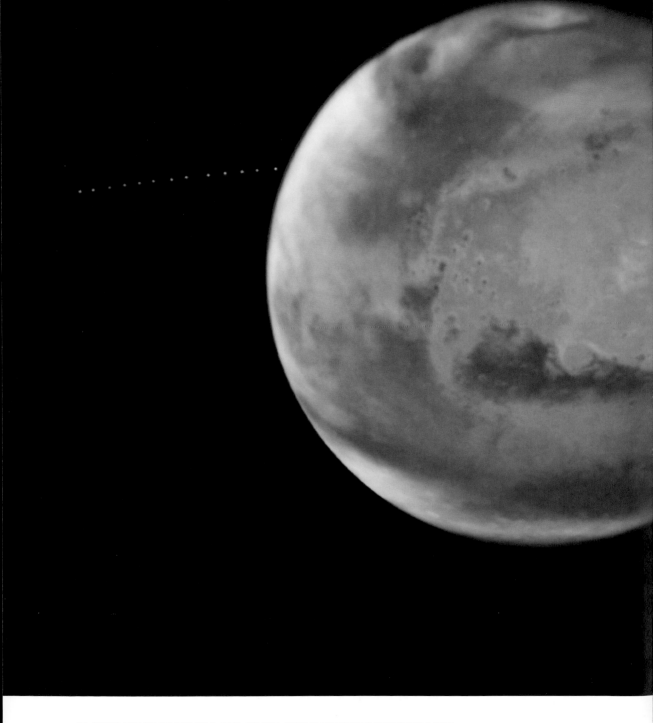

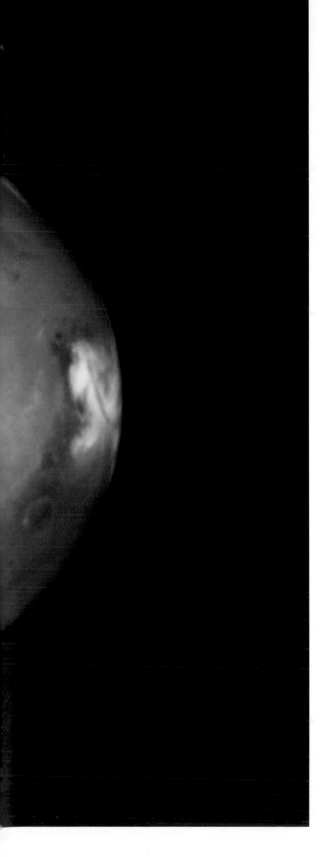

Mars and Tiny Phobos

The Hubble Space Telescope captured the tiny moon Phobos during its orbital 7 hour and 39 minute orbit around Mars. Hubble took thirteen individual exposures of Phobos in the course of twenty-two minutes that were then combined into one image.

Phobos is a small, elongated moon about sixteen miles at its longest dimension. Each individual exposure can be seen in this image to the left of the planet. It is the solar system's only satellite that completes its orbit in less time than its parent planet's day. The Martian day's length, 24 hours and 40 minutes, is nearly the duration of an Earth day.

Image credit: NASA, J. Bell (Cornell II.) and M. Wolff (SSI

Jupiter's Moons and Their Shadows

These two images show a procession of Jupiter's moons with their shadows crossing the planet on January 24, 2015. Three moons can be seen in the bottom image. Europa is on the lower left, Callisto is above (to the right of Europa), and Io is farthest east *(top right)* on Jupiter's sphere. The Great Red Spot Hubble recorded Jupiter's great red spot, which is a persistent zone of high pressure. This storm may be over 350 years old.

Ultraviolet Images of Comet Impacts

This ultraviolet image *(bottom)* of Jupiter was taken by the Hubble Space Telescope's Wide Field Camera. The image shows the many impacts by the fragmented Shoemaker-Levy 9 comet.

This photo was taken July 21, 1994 just hours after an impact— which created the third dark spot from the right on the bottom. The spots appear dark in ultraviolet because the dust kicked up into the stratosphere by the

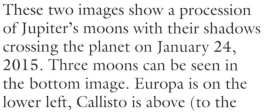

Image credits: NASA, ESA, and the Hubble Heritage Team (STScI/AURA)

Image credit: NASA, ESA, and A. Simon (Goddard Space Flight Center)

impact absorbs sunlight. Scientists will be able to track winds in the stratosphere by watching the evolution of these features. The dark spot above the planets equator is Jupiter's moon, Io. It is high above the planet.

Image credit: Hubble Space Telescope Comet Team

Image credit: Hubble Space Telescope Comet Team and NASA

Jupiter and the Colliding Comet

Impact sights from the collision with Comet Shoemaker-Levy 9 are visible in this Hubble photograph *(top)*. This image is a color composite of three filters at 9530, 550, and 4100.

Jupiter and Comet Shoemaker-Levy 9

This composite photo *(facing page top)* is compiled from separate images of Jupiter and the comet Shoemaker-Levy 9 in 1994. It was named after astronomers Carolyn and Eugene M. Shoemaker and David Levy who discovered it in 1993. The comet was observed to be orbiting Jupiter—not the Sun. Jupiter's tidal forces had fragmented the comet before its collision, and then in 1994 it collided with Jupiter.

The image *(top right)* of the approaching comet, shows 21 fragments (the lower half of the image), was taken on May 17, 1994. The portion of image showing Jupiter was taken on May 18, 1994. The dark spot on the planet is the shadow of the inner moon Io. The apparent angular size of Jupiter relative to the comet, and its angular separation from the comet when the images were taken, have been modified for illustrative purposes.

Europa, Jupiter's Moon

This image *(bottom)* was made by Hubble's STIS MAMA instrument using a filter sensitive for ultraviolet light. The focus is on the area above Europa's cold surface. It is thought that water had erupted from below an ocean (the seven o'clock position). It is possible that life-creating or sustaining conditions may exist in the heated oceans of this frozen world. Color was added to the grayscale image using blue color mapping. A separate grayscale image of Europa was superimposed with data from NASA Galileo and Voyager missions.

Image credit: NASA, ESA, STScI, and JPL

Image credit: NASA, ESA, W. Sparks (STScI), and the USGS Astrogeology Science Center

Saturn's Changing Seasons

This is a composite Hubble image taken with the Wide Field Planetary Camera 2. The image is of the planet Saturn as it travels around the Sun during a four-year period. Because of its tilt at 27 degrees, which is similar to Earth's tilt at 23 degrees, it experiences seasons. Saturn's year is equal to 29 Earth years. Although its day is only 10 hours long. In this series, it is moving from autumn to winter in its Northern Hemisphere. Because of the tilt, its orientation toward the Sun changes, causing the different seasons.

By the last image in the sequence on the upper right, the northern hemisphere is in its winter solstice while the southern hemisphere is in its summer solstice.

Astronomers are investigating the details and variations in the color and brightness of the rings to find out more about them. How were

Image credit: NASA and The Hubble Heritage Team (STScI/AURA) Acknowledgment: R.G. French (Wellesley College), J. Cuzzi (NASA/Ames), L. Dones (SwRI), and J. Lissauer (NASA/Ames)

they formed? How are they changing with time, and with the seasons? The rings are made of dust and water ice that sometimes collide as they orbit around their planet. Do the seasons affect this? Saturn's gravity keeps the debris spread out so it doesn't form a moon or larger pieces.

The slight pale reddish tint to the rings is caused by the presence of organic material mixed with the water ice. There are visible clouds, and the cloud band color variations are upper atmosphere smog made when ulta-viot radiation from the Sun interacts with the methane gas. In some places the gases are hotter and denser. This planet is a gas giant. There is no solid surface on Saturn.

Saturn's Rings

As noted, images taken with different instruments are often used together or combined into a single image. The image of Saturn *(above)* was taken with the Wide Field Planetary Camera 2 onboard Hubble. The detail im-

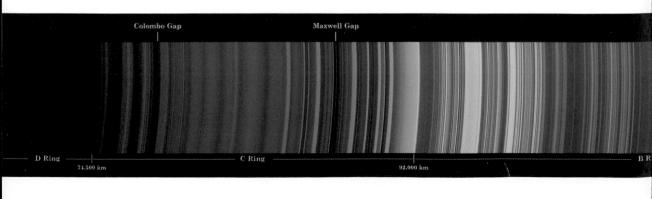

Colombo Gap   Maxwell Gap

D Ring   C Ring   B R

74,500 km   92,000 km

Image credit: NASA and
The Hubble Heritage Team (STScI/AURA)

age below of the rings and their gaps
was taken by the ISS-Narrow Angle
Instrument on the Cassini Orbiter.

Image credit: NASA/JPL/Space Science Institute

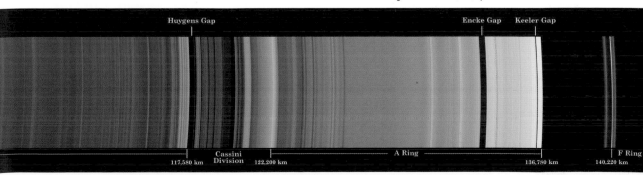

Huygens Gap   Encke Gap   Keeler Gap

Cassini Division   A Ring   F Ring

117,580 km   122,200 km   136,780 km   140,220 km

Image credit: NASA/JPL/STScI

Uranus

This NASA Hubble Space Telescope image shows Uranus has four major rings, and it shows ten of its seventeen known satellites. Hubble record-ed bright clouds on Uranus. Uranus is tilted and appears to orbit the Sun on its side. Uranus is 1.6 billion miles (2.6 billion kilometers). Uranus and Earth at their closest are 1.6 billion miles away from each other.

Bright Clouds on Uranus

Image credit: NASA/JPL/STScII

These three NASA Hubble Space Telescope images *(top right)* of the planet Uranus show the motion of a pair of bright clouds in the planet's southern hemisphere and a high altitude haze that forms above the planet's south pole.

The great distance—more than a billion miles—and the substantial magnification of these Uranus photographs creates images that are pixelated, but nonetheless, informative.

Image credit: ESA/Hubble and NASA, L. Lamy/Observatoire de Paris

Aurorae on Uranus

This is a composite of images taken by Voyager 2 and two observations made by Hubble—one for the ring and one for the aurorae. This image records two shocks caused by solar wind bursts that produced aurorae on the planet Uranus. There is evidence that the aurorae rotate with the planet. As a result of these observations, the planet's magnetic poles were re-assessed.

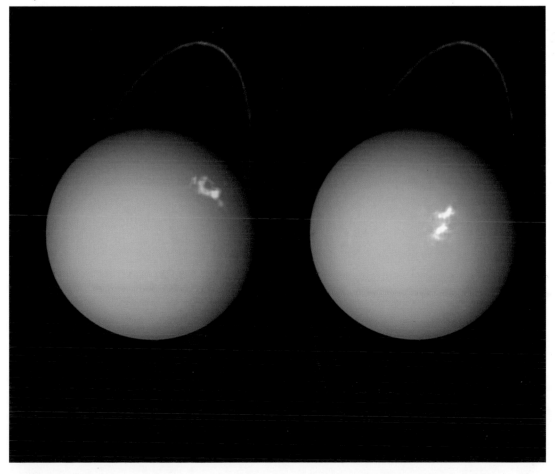

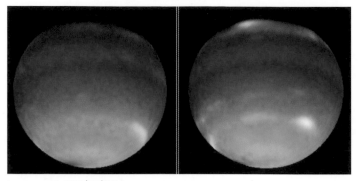

Image credit: NASA/JPL/STScI

Neptune's Weather

Neptune is 2.7 billion miles from Earth. Hubble's Wide Field Planetary Camera 2 took these two composite images *(left)* of Neptune on August 13, 1996, during its 16.11-hour rotation. A team, directed by Lawrence Sromovsky of the University of Wisconsin-Madison's Space Science and Engineering Center, made these observation, that are a combination of

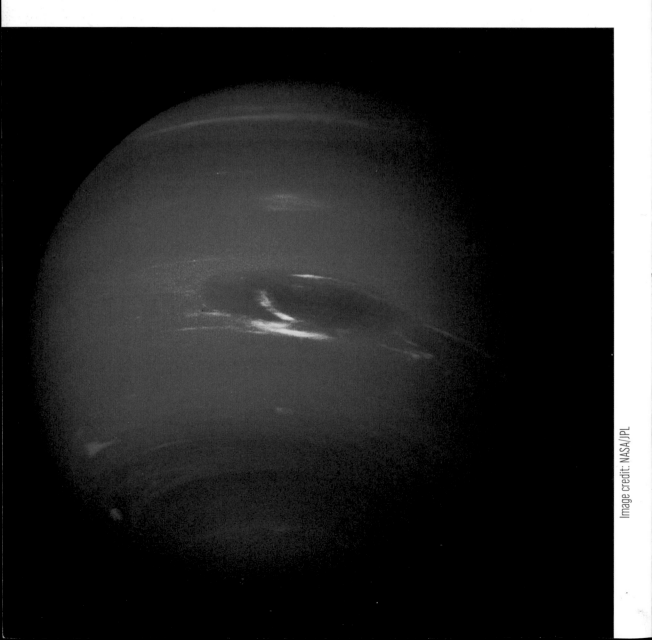

Image credit: NASA/JPL

wavelengths that bring out Neptune's weather features. Neptune is mainly blue because red and infrared light is being absorbed by its methane atmosphere. Clouds are above most of the methane, so they appear white, and the very highest clouds are yellow, and red, which can be seen on the top of the planet. It's estimated that Neptune's equitorial winds are almost 900 miles per hour—indicated by a dark blue belt. The ring near the bottom of the planet is an area that absorbs blue light and so it appears green.

Neptune has a highly developed weather pattern. This image *(facing page bottom)* was taken by Voyager 2, when it was 4.4 million miles from the planet. One reason why the image is larger than other Hubble images of Neptune is because Voyager 2 was able to get much closer to Neptune—over two billion miles closer. It will take Hubble, Voyager 2, and future mission images and data, to unravel Neptune's secrets.

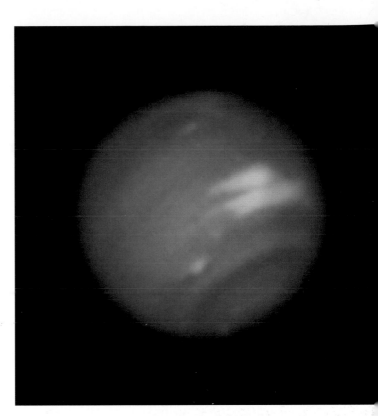

Image credits: NASA, ESA, and M.H. Wong and J. Tollefson (UC Berkeley)

Neptune's Storm

This image *(top right)* taken in May 2016 affirms the presence of a vortex in the atmosphere of Neptune which was first spotted in September 2015 by September 2015, the Outer Planet Atmospheres Legacy (OPAL) program. This identifies these types of storms as fairly long lived events. This vortex, seen here in magnification *(bottom)*, is about 3000 miles across.

Pluto and Its Moons: Charon, Nix, and Hydra

In 2005 NASA's Hubble Space Telescope discovered a pair of small moons orbiting Pluto: Nix and Hydra. In this image *(facing page)*, each of the satellites in this pair is smaller and farther away from Pluto and its larger moon, Charon. Some consider Pluto and Charon double planets or binary planets because they are seen jointly as a planetary mass revolving around each other. Looking at their similar size and proximity in this image, the statement about the planet set is understandable.

Pluto is very far away compared to the planets in the solar system, and it is also very small. So this is what we can see with Hubble. Pluto's images also appear small, with fewer details compared to the images of stars and galaxies because these other objects, although very far away, have two things Pluto doesn't have. First, Pluto does not emit light like a star or supernova does; it reflects the light from our star, the Sun. So it is less bright. The tiny dots of Nix and Hydra show us that they are even fainter than Pluto—about 5,000 times fainter. Second, Pluto is very small compared to a galaxy, which can be many light-years across—that's more than a trillion times bigger than Pluto.

This close up image of Pluto *(below)* showing amazing detail is from the New Horizons interplanetary space probe. Pluto is so far away, it took over nine years to travel to the distant dwarf planet.

Image credit: NASA/Johns Hopkins University Applied Physics Laboratory/Southwest Research Institute/Lunar and Planetary Institute

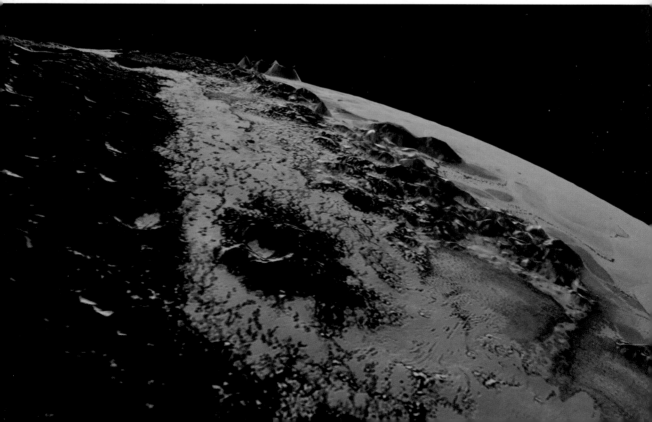

Image credit: NASA, ESA, H. Weaver (JHUAPL), A. Stern (SwRI), and the HST Pluto Companion Search Team

Pluto's Tumbling Moons

The Hubble Space Telescope has also discovered four tiny satellites orbiting Pluto and Charon. It also found that two moons, Hydra and Nix, are tumbling because of Pluto and Charon's gravitational pull as they rotate around each other.

Our Solar System's Comets

Our solar system is home to comets, whose eccentric orbits bring them from the outer reaches of the system to very close to the Sun. While near the Sun, they become more visible, with a tail projecting in the opposite direction of the Sun.

Comet 332P/Ikeya-Murakami *(left)* is a fragmenting comet. This is a sequence of images *(middle left)* showing comet 252P/LINEAR as it passed by Earth. It has been questioned whether P/2010 A2 is or is not a comet. Its tail is split suggesting it had a highspeed collision with another object.

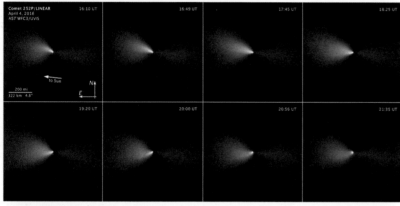

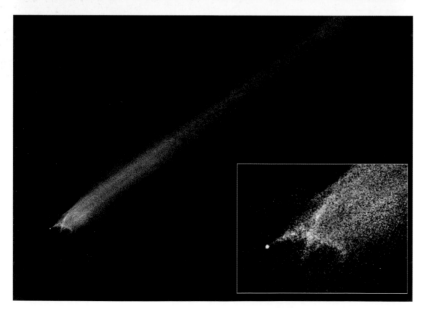

Image credit (top): NASA, ESA, and the Hubble Heritage Team (STScI/AURA)
Image credit (middle): NASA, ESA, and J.-Y. Li (Planetary Science Institute)
Image credit (bottom): NASA, ESA, D. Jewitt (UCLA)

Image credit: NASA, ESA,
and the Hubble Heritage Team (STScI/AURA)

Sungrazing Comet

Sungrazing comets come very close to the Sun. Outgassing can happen as it gets closer to the Sun making it more visible. Comet ISON is approaching the Sun. This image is a composite of several exposures; one is a red and yellow-green light image of the stars and galaxies in the background. The other is a black & white exposure of the comet. This allowed a long enough exposure for the background stars and galaxies' light to register. It also allowed the comet to be in focus and reduce motion blur.

Stars, Supernovas & Planetary Nebula

A star is a luminous sphere of plasma. Stars are held together by their own gravity. We can see stars because of thermonuclear fusion. Each phase of a star's life and its circumstance projects a characteristic electromagnetic radiation signature.

Supernovas and planetary nebulas occur when stars are of a particular size and grow older. The amount of visible energy given off from stars that reach the end of their lives makes them observable from very far away. They are prime targets for the NASA Hubble Telescope because of their brightness.

Image credit: NASA/ESA/P. Jeffries (STScI)

N6946-BH1 Became a Black Hole

N6946-BH1, 25 times as massive as our Sun, brightened in 2009 then disappeared. Scientists making observations with the Large Binocular Telescope, NASA's Hubble, and Spitzer space telescopes concluded the star must have become a black hole. Normally a supermassive star becomes a supernova, which would be very visible. N6946-BH1 appears to have imploded instead as illustrated in the Hubble-informed illustration below.

Image Credit: NASA/JPL-Caltech/STScI/CXC/SAO

PIA03519:
Cassiopeia A

PIA03519, also called Cassiopeia A, is a supernova remnant, located in the constellation Cassiopeia, 10,000 light-years away. The supernova was visible in the our sky about 320 years ago. A barely detectable neutron star is all that remains. This image is a false-color image combined from Hubble and two other telescopes of NASA. Spitzer Space Telescope's infrared data is red; Hubble's visible data is yellow, and Chandra X-ray's data is green and blue.

Signs of Star Formation

A glowing nebula is often the most evident sign that new stars are being born. This image of debris, captured by Hubble's Wide Field Planetary Camera 2, is from an exploded massive star in the Large Magellanic Cloud galaxy. Debris from exploded supernovae often become star nurseries.

Ring Nebula (M57)

This 1998 Hubble image, created with the Wide Field Planetary Camera 2, looks through the center of a dying star. There are long clumps of material lodged in the nebula at its edge. The image is compiled from three black & white images, and then each were assigned a color. The blue center, which indicates hot helium, is around the dying star at the middle. Green shows ionized oxygen, and red shows ionized nitrogen, which come from the cooler gases. These gases are illuminated by the ultraviolet radiation coming from the bright, dying star.

Image credit: NASA/JPL- Caltech/ESA, and the Hubble Heritage Team (STScI/AURA)

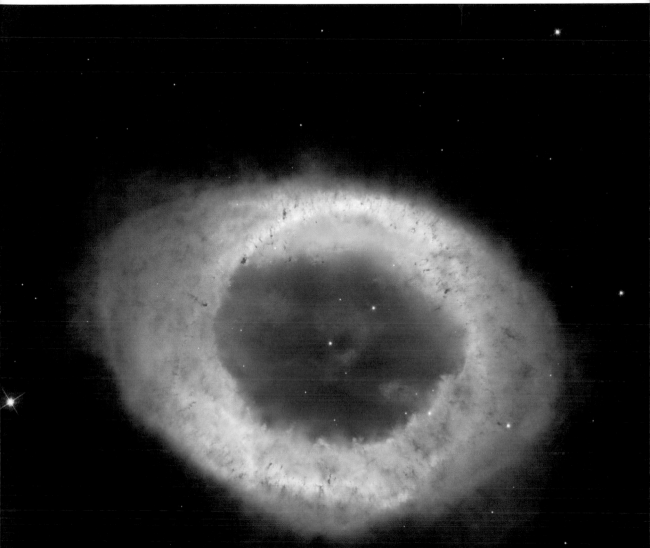

Stars and Globules

This image of IC 2944, taken in 1999 by Hubble's Wide Field and Planetary Camera 2, is of globules (the dark clouds) and bright stars in a star-forming region in the constellation Centaurus. These globules, discovered by astronomer A. D. Thackeray, are often linked to hydrogen-emitting, star-formation regions. Two clouds appear to be overlapping. These globules seem to be part of the process of forming new stars. IC 2944 is comparatively close to us at about 5,900 light-years away.

Image credit: NASA and The Hubble Heritage Team, NASA, and The Hubble Heritage Team (STScI/AURA) Acknowledgment: Bo Reipurth (University of Hawaii) (STScI/AURA)

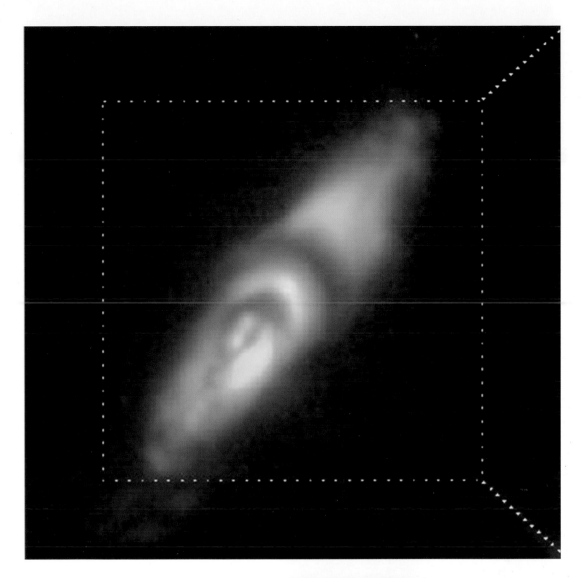

Protoplanetary Nebulas

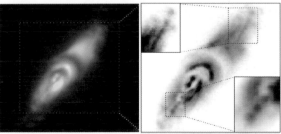

Image credits: NASA/JPL

Characteristic of this protoplanetary phase of a star's life, the outer layer is expelled resulting in a cloud that is lit up by ultraviolet light from inside the star. Protoplanetary nebulas are rarely seen because they are a short-lived stage before a star becomes a planetary nebula. Images taken of these infrequent opportunities help to study this phase of star evolution.

Image credit: NASA/JPL/STScI/AURA

One of the Hottest White Dwarfs

NGC 2440 a planetary nebula, with it's white dwarf star, is one of the hottest known, with a surface temperature of almost 400,000 degrees Fahrenheit (200,000 degrees Celsius). The non-circular shape suggests that it ejected its mass periodically in different directions.

In this image, the blue shows high concentrations of helium; the blue-green shows the oxygen; and the red shows the nitrogen and hydrogen.

Cat's Eye Nebula: Beautiful Dust Shells

This is a detailed view of NGC 6543, informally called the Cat's Eye Nebula. It is a complex nebulae and one of the first planetary nebulas seen. It has concentric gas shells, jets of high-speed gas, and shock-induced knots of gas. Based on observations, it's suggested that the star ejected a large about of matter in a series of regularly timed intervals creating these dust shells that share the same center.

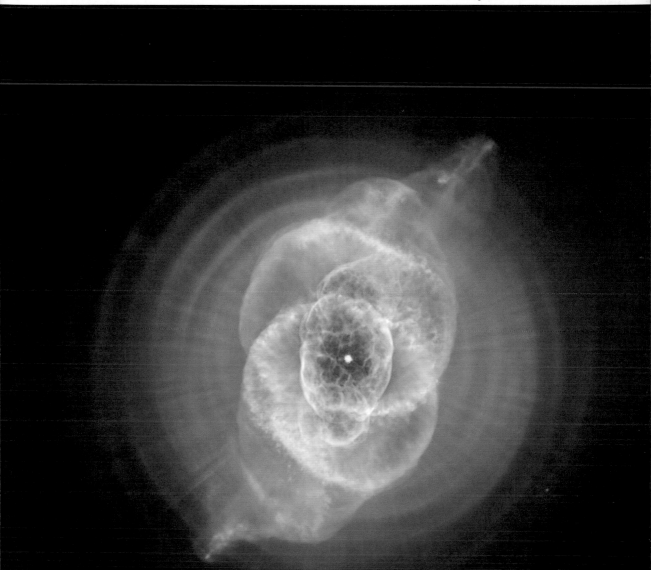

Helix Nebula: PIA03678

An Earth-size white dwarf star sits in the middle of this planetary nebula, emitting massive amounts of hot gas and ultraviolet radiation. Combining data from NASA's Hubble (visible light data) and Spitzer Space (infrared data) telescope, a false-color image has been made. The center blue area is the hottest, yellow is hot, and red is warm. It's believed the red areas are somehow shielded from the ultraviolet radiation, allowing these zones to stay cooler.

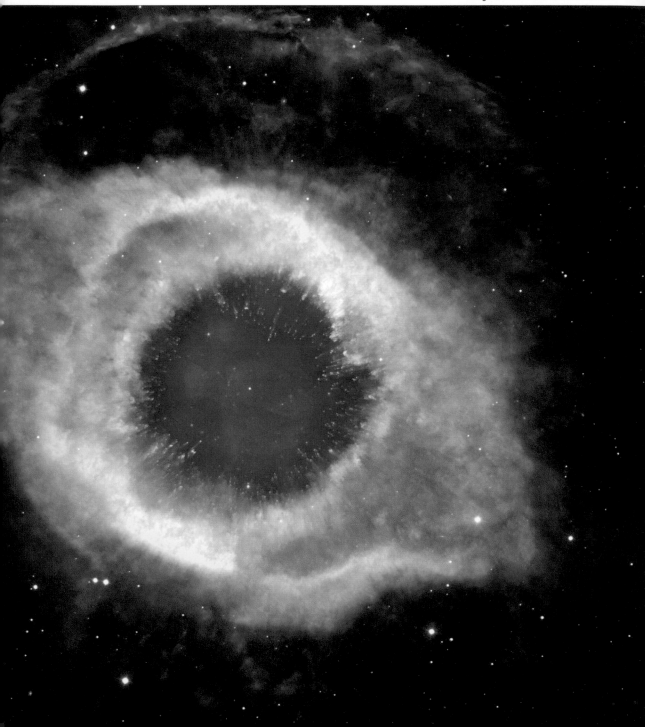

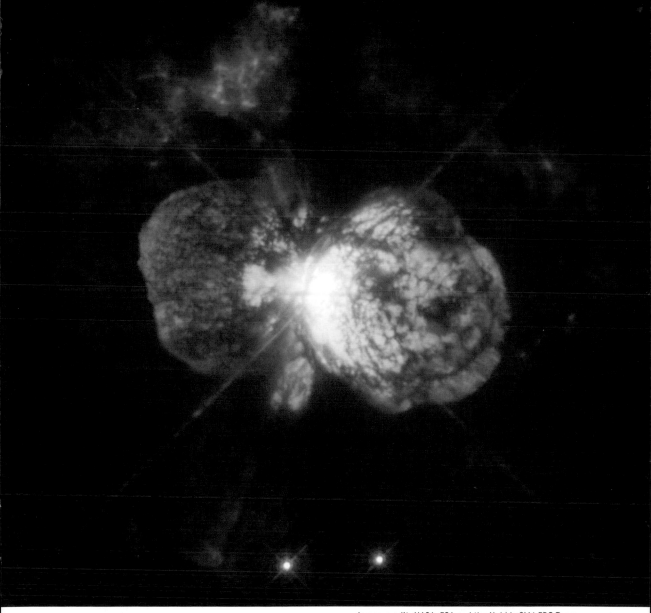

Image credit: NASA, ESA and the Hubble SM4 ERO Team

Eta Carinae: Twin Stars

This image taken by the Space Telescope Imaging Spectrograph (STIS) on the NASA's Hubble Space Telescope shows a pair of massive stars—twins. One is more massive than the other. The image shows streams of charged particles from their eruptions. These eruptions have been seen over the last 200 years. At one time it could be seen with the naked eye. It faded to invisible by 1858, reappearing again in the 1990s, and then, its brightness doubled around 1998 and 1999.

Image credit: ESA/Hubble & NASA, Acknowledgement: Judy Schmidt

Death of a Star: Calabash Nebula

The Calabash Nebula is also known as the Rotten Egg Nebula because it contains lots of sulphur, which can smell like rotten eggs. This nebula is transforming from a red giant to a planetary nebula as it nears its death. The star is emitting its outer layers of gas and dust into the space that surrounds it. It is in a relatively short phase called a protoplanetary nebula before it becomes a planetary nebula.

The Crab Pulsar and Supernova Remnants

The neutron (right-most star of the two brightest stars in the center) star in the middle of this nebula is the Crab Pulsar and has the same mass as our Sun. However, it is only a few miles in diameter—much smaller than our Sun. It was first observed by the Chinese in 1054 A.D. and can be seen through amateur telescopes.

It sends out pulses of radiation and charged particles, some which look like threads. It spins thirty times a second. It has a powerful magnetic field, which is the remnant of supernova SN 1054. It is called a pulsar because from Earth's vantage point, as it rotates its electromagnetic radiation emissions appear to rapidly pulsate. Astronomers can use pulsars like this one to study gravitational radiation.

Credit: NASA and ESA

Black Holes and Quasars

Image Credit (facing page): NASA/JPL-Caltech/Roma Tre Univ.

Image credit: NASA, ESA, the Hubble Heritage Team (STScI/AURA), and R. Gendler (for the Hubble Heritage Team)

Quasars are typically extremely massive stars that emit large amounts of energy. There is evidence that each quasar contains a massive black hole, and that it's a stage in the evolution of many large galaxies.

Difficult to See

This image *(facing page)* of Galaxy NGC 1068 was composed of visible light from NASA's Hubble Space Telescope, and high X-rays (magenta) from NASA's Nuclear Spectroscopic Telescope Array, or NuSTAR. The X-ray is from an active supermassive black hole in the center of the galaxy. NGC 1068 is relatively close to our galaxy—about 47 million light-years away.

Fueling a Black Hole

Gases from the galaxy M106 are thought to be falling into and fueling a supermassive black hole at its center.

Image credit: NASA/JPL- Caltech/GSFC

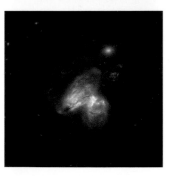

Super Massive Black Hole

This image *(above)* shows two galaxies in relatively close proximity, and is a combination of the two lower images. Each galaxy has a massive black hole in its center. The portion of the image coming from the NSTAR *(middle left)* shows that the black hole in the galaxy on the right is the most active in accumulating additional matter. The bottom image shows the colors from red to green to blue show increasing amounts of kiloelectron volts. The bottom image shows visible light only.

Obscured, Active Galactic Nucleus

IC 3639 is a galaxy with an active center. This image combines data from the Hubble Space Telescope and the European Southern Observatory. Hidden by gas and dust is its supermassive black hole. The galaxy's active galactic nucleus was confirmed by NuSTAR, NASA's Chandra X-Ray Observatory and the Japanese-led Su-zaku satellite. The galaxy's nucleus's brightness is obscured as is the presumed supermassive black hole at its center. This galaxy is thought to be 170 million light years away.

Scientists have calculated how much dust is obscuring the center of this galaxy. They have determined using NuSTAR, a space-based X-ray telescope, to measure the higher energy X-rays that the galaxy center is obscured and much more luminous than we see.

Image credit: NASA/JPL- Caltech/ESO/STScI

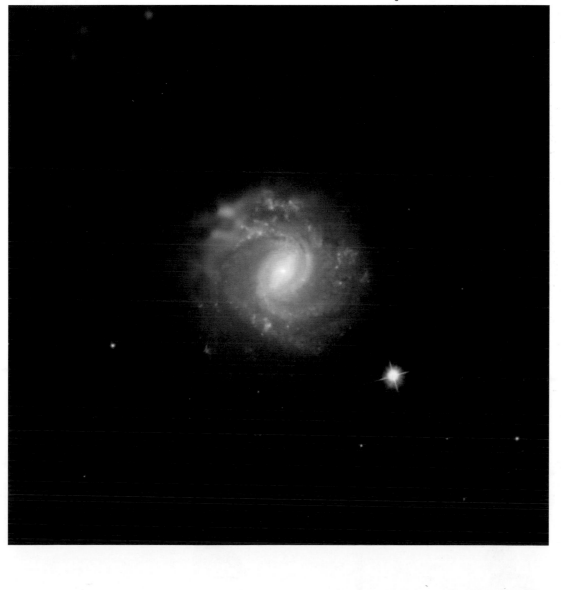

Dark Matter
and Dark Energy

Image credit (facing page): Credit: NASA, ESA, CXC, NRAO/AUI/ NSF, STScI, R. van Weeren (Harvard-Smithsonian Center for Astrophysics), and G. Ogrean (Stanford University)

Image credit (top): NASA, ESA, D. Harvey (École Polytechnique Fédérale de Lausanne, Switzerland), R. Massey (Durham University, UK), the Hubble SM4 ERO Team, ST- ECF, ESO, D. Coe (STScI), J. Merten (Heidelberg/Bologna), HST Frontier Fields, Harald Ebeling (University of Hawaii at Manoa), Jean-Paul Kneib (LAM), and Johan Richard (Caltech, USA)

Many things in space are invisible to us because of their distance and wavelength. Dark matter and dark energy escape our visual sense and our instruments' detection. Our knowledge is based on calculations of observable matter and its interactions with other objects and within galaxy clusters. However, telescopes like Hubble can enhance images by including non-visible data and putting it into a visible spectrum.

Dark Matter
in Galaxy Clusters

These images show galaxy clusters and the dark matter in them. Scientists suggest that the universe is made up of about 68 percent dark energy, about 27 percent dark matter, and about 5 percent normal matter. Dark matter and normal matter are held together by gravitational forces. These gravitational calculations are the main reason why we know dark matter exists. It does not emit or absorb light. Several ideas, such as cold dark matter and fuzzy dark matter, are being studied to help explain this hidden material. We know it is there and has to be accounted for. The nature of dark matter remains a mystery.

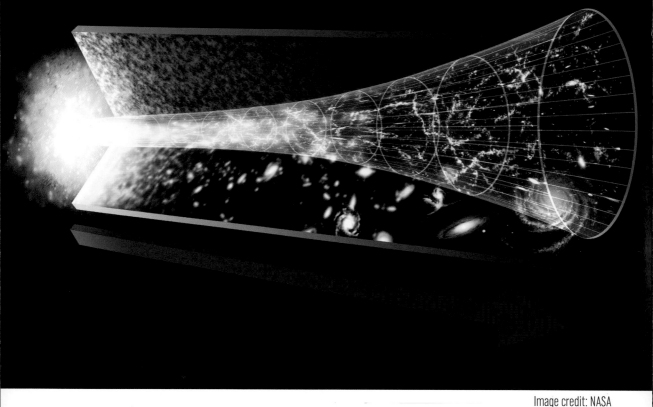

Image credit: NASA

Mysterious Dark Energy

Scientists think the mysterious dark energy is a form of energy that may help explain why the universe is expanding at an accelerated rate. The astronomer Edwin Hubble discovered almost 100 years ago that the universe is expanding. Since NASA's Hubble Telescope observations in the 1990s, scientists have found that the rate is accelerating.

According to a current model of cosmology, dark energy makes up the majority of energy in the universe. The mass and energy of dark matter account for the next largest portion, and ordinary matter with a small amount of other components, such as neutrinos and photons accounts for the last six percent. Although the concentration of dark energy is very low, its force dominates the universe because it is uniformly across all of space.

Dark matter and ordinary matter are gravitationally responsive. Dark energy has the opposite effect; it appears to be a repelling force. Some astronomers believe this force may be growing in strength.

A team led by Nobel Laureate Adam Riess began to measure the universe's expansion rate, known as the Hubble constant. They are refining this measurement and uncovering more about dark energy.

About 13.7 billion years ago the universe began from a very dense and hot state. It expanded rapidly eventually forming stars and galaxies. According to scientists, the rate of expansion after some time slowed, but at one point it began to speed up again. The Big Bang theory is being refined again and again as more of its mystery unravels.

Image credit: NASA

Measuring Dark Energy

Using gravitational lensing *(top)* to reach far back into the earlier universe, scientists can see distant and early supernovae. The supernovae event's light took so long to get to us, we are actually seeing how it looked billions of years ago.

Five supernovae images *(bottom)* are in the top row. In the row below are their host galaxies before or after the supernovae event. The supernovae can be used to measure the expansion rate and how it is affected by dark energy.

Image credit: NASA, ESA, and A. Riess (STScI)

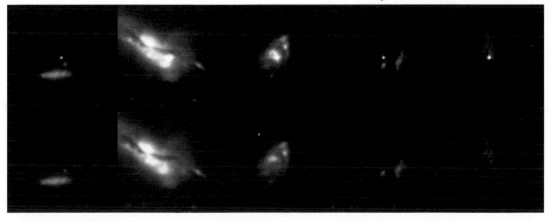

Our Galaxy
The Milky Way

Image credit: ESA/Hubble and NASA
Acknowledgement: Judy Schmidt (Geckzilla)
Image credit (facing page): NASA/ESA

The Arches Star Cluster

Our solar system is within the Milky Way galaxy. Edwin Powell Hubble discovered that there were galaxies beyond the Milky Way. The Milky Way is a barred spiral galaxy. Its diameter is somewhere between 100,000 and 180,000 light-years. Our galaxy is estimated to have between 100 and 400 billion stars and an estimated 100 billion planets. The center most likely has a super-massive black hole, which is marked by an intense radio source, named Sagittarius A*.

The Arches cluster *(facing page)* is the densest known star cluster in our galaxy. It is a fairly young astronomical object (between two and four million years old) and is located in the center of the Milky Way.

Hen 2-437

Hen 2-437 *(above)* is a planetary nebula. There are thousands of planetary nebula within the Milky Way. Hen 2-437 is a bipolar nebula; the aging star's material has ejected its material in opposite directions, creating two lobes.

Image credit: ESA/Hubble and NASA

Our Neighbor, Super Star Cluster Westerlund 1

This is Westerlund 1, a young super star cluster determined to be 15,000 light-years away in our Milky Way neighborhood. It contains one of the largest stars ever discovered, Westerlund 1-26, which is classified as a red supergiant. Most of the stars in this cluster are young stars of about 3 million years old, compared to the Sun, which is about 4.6 billion years old.

A Filament of SN 1006

This is a very thin ribbon of a supernova remnant. A star referred to as SN 1006 exploded in the Milky Way galaxy more than 1,000 years ago. In 1006 A.D. people witnessed this bright supernova, which stayed visible for over two years. At its brightest, it would have been visible in the daytime sky.

The supernova would have happened far off from the lower right corner of the image (outside the image frame). The motion would be toward the direction of the upper left. It is currently expanding at tremendous speeds—about six million miles per hour. This is barely noticeable to us because of the vast amount of distance.

Image credit: NASA, ESA, and the Hubble Heritage Team (STScI/AURA)
Acknowledgment: W. Blair Johns Hopkins University

Image credit: NASA, ESA, and G. Brammer

Sagittarius A: Our Own Galaxy's Black Hole

Sagittarius A is the black hole at the heart of our galaxy, the Milky Way. Looking at this image, it is easily understood why our galaxy is called the *Milky Way*—stars and clouds of dust are so concentrated, it does look like milk. However, the galaxy's central supermassive black hole is hidden. It is in the center of this image. Sagittarius A has an enormous gravitational pull. We know it is there because scientists have observed large quantities of stars revolving around it.

Gabriel Brammer, a fellow at the European Southern Observatory and an ESO photo ambassador posted this image which uses infrared Hubble data collected from 2011.

Image credit: NASA, ESA, and G. Brammer

A Closeup

This image is a detail of the the invisible Sagittarius A*.

Stars of the

Milky Way

These two images were taken by the Hubble and show the thick endless mass of stars in our view of the Milky Way.

Image credits: NASA, ESA, and G. Brammer

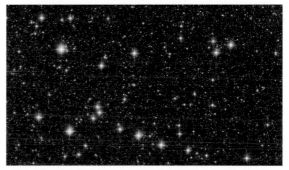

The Milky Way

Using Three Telescopes

This image *(below)* is a combination of observations using infrared light, X-ray light, and visible light, which reveal the activity beyond the dust. Each image *(facing page)* shows dif-ferent wavelength views of the galactic center region. Spitzer's infra-red-light records the stars. Hubble uncovered many more massive stars across the region. Chandra records X-rays indicated by pink (lower energy X-rays) and blue (higher energy). A massive black hole is indicated by the bright region in the lower right. The diffused X-ray light comes from gas heated to millions of degrees by

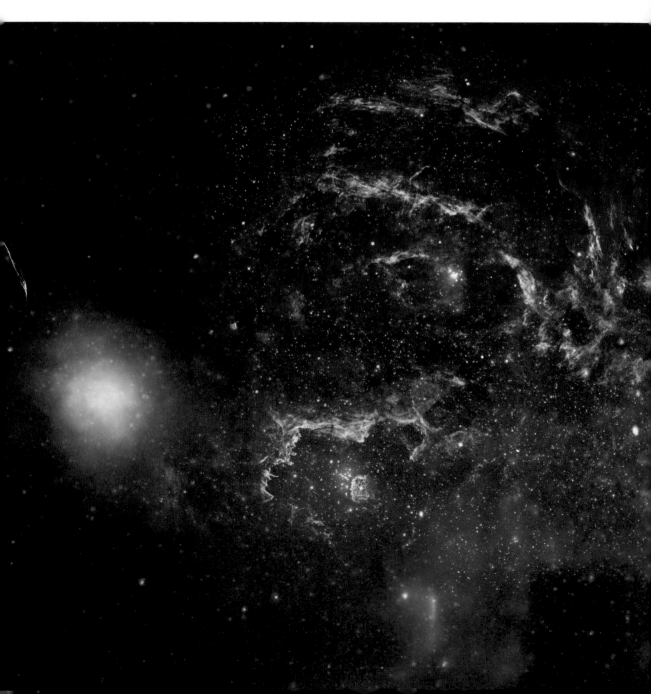

outflows from the supermassive black hole, winds from giant stars, and stellar explosions. This dynamic area can be viewed as a central region in our Milky Way galaxy.

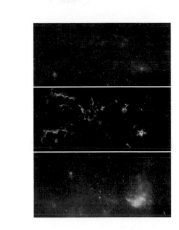

SPITZER • INFRARED

HUBBLE • VISIBLE

CHANDRA • X-RAY

Image credit: NASA/JPL-Caltech/ESA/CXC/STScI

Image credit: NASA/STScI

Milky Way Stars: Birth and Death

This sparkling star cluster taken by NASA's Hubble Space Telescope contains some of the brightest stars seen in the Milky Way galaxy. The cluster is a collection of young massive, luminous stars, only 500,000 years old. They burn their hydrogen fuel and will result in supernova blasts. The eventual outcome will be a new generation of stars.

This image of Trumpler 14 was made with Hubble's Advanced Camera for Surveys. Blue, visible, infrared broadband filters, and filters that isolate hydrogen and nitrogen from the gas surrounding the cluster were used to assemble the image.

Herbig-Haro 24, HH 24

Herbig-Haro 24 has small patches of nebulosity linked to newly born stars. Narrow jets of gas are ejected, running into the surrounding clouds of gas and dust at great speeds. Later, these clouds of gas and dust will gather around the rotating star.

These knotty clumps of nebulosity are jointly known as Herbig-Haro (HH) objects. They are typical in star-forming regions—this one being in the Milky Way's Orion B molecular cloud complex. Eventually, the surrounding material will collapse into a disk under its gravity around the new star. Planets can form in this flattened disc.

Image credit: NASA and ESA

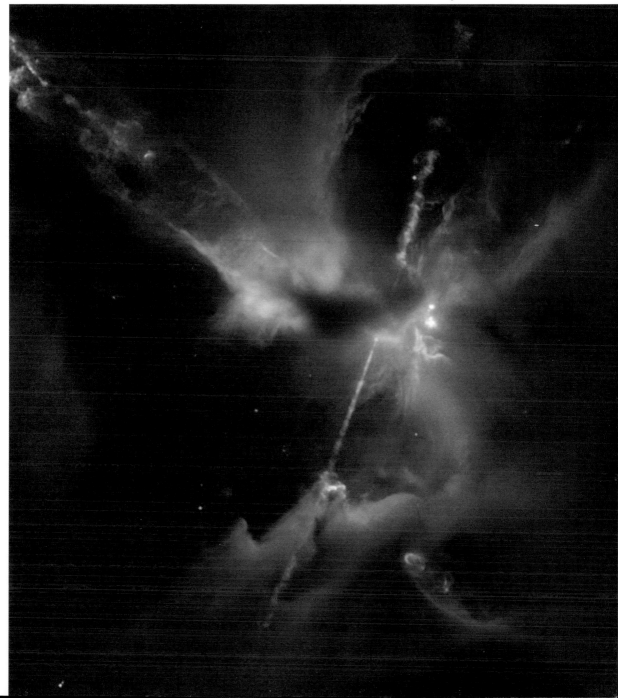

Pillars of Creation

This is an image of part of the Eagle Nebula. It was originally imaged in 1995 *(bottom right)*, and again in 2011. This higher resolution image *(above)* was taken in 2014. The Eagle Nebula, also called the Star Queen Nebula and the Spire, holds several active star-forming gas and dust regions. The Pillars of Creation is one of these regions. The nebula is also catalogued as Messier 16 and NGC 6611. The Eagle Nebula and IC 4703 are parts of a diffuse, emission nebula. It is an active star forming region almost 7,000 light-years away from the Sun. The nebula is located on an inner spiral arm of the Milky Way next to our arm of the galaxy.

Tower of Gas

This image *(top right)* is a small section of the Eagle Nebula. It is 57 trillion miles long (91.7 trillion km).

A Stair Step Image

This is the original image *(bottom right)* of the nebula made in 1995. The original Hubble instrument WFPC2 had four cameras: three wide-field cameras and one high-resolution planetary camera. At the expense of creating a higher-resolution image, the high-resolution planetary camera created a smaller image. This part of the image when stitched to the other three has a black, blank area—the upper right portion of the image. It's as if this one corner has a telephoto lens instead of another wide-angle lens.

Image credits (left and right pages) : NASA, ESA, and the Hubble Heritage Team (STScI/AURA)

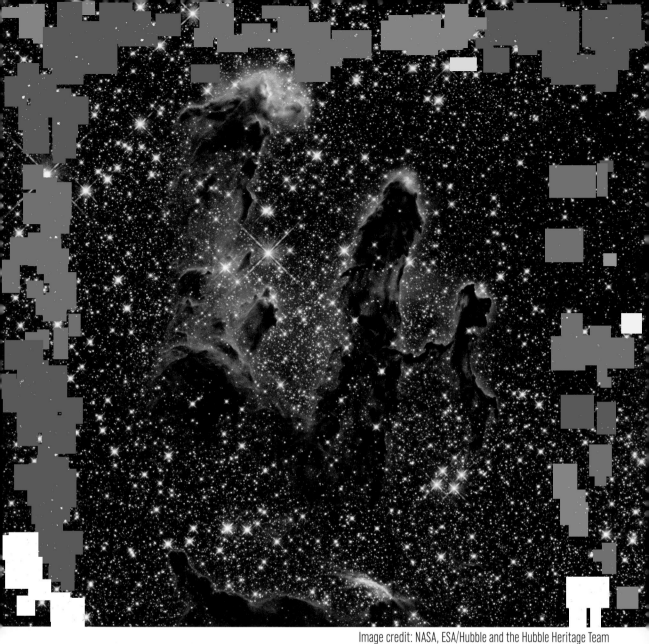

Image credit: NASA, ESA/Hubble and the Hubble Heritage Team

Infrared Shows Brilliant Stars

This view of the Pillars of Creation is seen with infrared light. The dust and cloud do not obscure the stars that pepper this area of the sky. The trans-parent outlines of the Pillars suggest their presence without erasing the countless brilliant and infant stars that are within and behind the nebula.

Image credit: NASA/ESA and H. Richer

Old Stars in the Milky Way

Globular Cluster M4 contains several hundred thousand stars. This star cluster is located in our Milky Way, and contains some of the oldest burned-out dying stars in our galaxy. Classified as white dwarfs, they are about 12 to 13 billion years old, nearly the age of the universe.

Image credit: NASA/ESA/Hubble Heritage Team (STScI/AURA)

Horsehead Nebula

The Horsehead Nebula also known as Barnard 33 and IC 434. It is around 1,500 light-years from Earth within the Milky Way galaxy. It is one of the closer celestial objects that can be seen with ground telescopes.

The Horsehead Nebula is a favorite object for amateur and professional photographers because of its appearance as a figure covered in transparencies, rising from the shadows beneath. The infrared camera data allows stars to show through while its surroundings appears to bubble and foam.

Hubble images of nearby objects inspire ground-based telescope viewers to see theses scenes with their own eyes and captures them with their own cameras *(top right)*.

Image credit: Ken Crawford

Image credit: ESO

The Orion Nebula

Image credit: NASA, ESA,
M. Robberto (Space Telescope Science Institute/ESA)
and the Hubble Space Telescope Orion Treasury Project Team

The Orion Nebula is the star-forming region closest to our solar system, and is within the Milky Way. It is classified as a H II region. Over 3,000 stars are in this image. The ultraviolet radia- tion of several young massive stars are carving out an area in the central re- gion. Over 520 Hubble images taken between 2004 and 2005 were used to make this final image.

Image credit: NASA, ESA, M. Livio and the Hubble 20th Anniversary Team (STScI)

Carina Nebula

This pillar of gas and dust is three light-years tall. There are infant stars inside the pillars. This stellar nursery is located 7,500 light-years away. The pillar is shaped by streams of charged particles. Ionized gas and dust flow from the structures. The lower, denser parts have resisted being eroded.

Gas shoots in opposite directions at the top of the image. Another pair of jets can be seen. These jets are known as HH 901 and HH 902, and are an indication of new star birth with swirling gas and dust discs circling the infant stars.

The blue in this image indicates oxygen, the green indicates hydrogen and nitrogen, and the red indicates sulphur.

Image credit: NASA/Space Telescope Science Institute

Space is filled with galaxies. Galaxies are made of stars, stellar remnants, interstellar gas, dust, and dark matter that is gravitationally bound. Galaxies can be dwarfs with only a couple hundred million stars or giants with one hundred trillion stars or more. Each orbits its center of mass, which most often is a black hole.

Edwin Hubble first classified galaxies into these categories: elliptical, normal spiral, barred spiral (like our Milky Way), and irregular. It is thought that spiral galaxies tend to evolve into elliptical galaxies by merging with other galaxies.

As of 2016, estimates of how many galaxies there are in the universe place the number at around two trillion. As galaxies evolve, they tend to become larger as a result of their merging with other galaxies in their maturing process.

A Twisted Disk

ESO 510-G13 (above) is a spiral galaxy, like our Milky Way. It is on edge, so it looks flat. It has been speculated that this slightly twisted galaxy collided with another causing the warping in its disk shape.

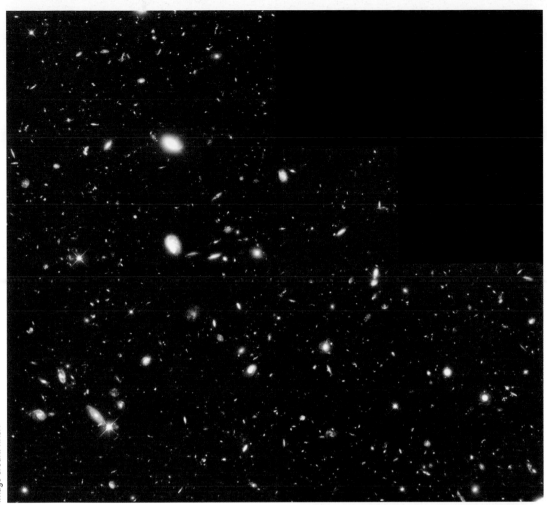

Image credit: NASA

Hubble Deep Field

This image taken in 1995 was the
deepest ever taken of the universe at
the time. It contains thousands of
nearby and distant galaxies. It is only
a tiny sample of our universe as can
be seen in the outline of the section
of the sky where it was taken (bottom right). This image was important
to scientists because it gave them
a look at very young galaxies and
their evolution.

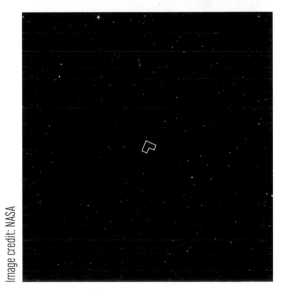

Image credit: NASA

Ultra-Deep Field

This image taken in June 2014 includes the most distant galaxies ever seen with an optical telescope. This magnification and the increased distance was made possible with several upgrades in equipment and software that Hubble received. Amazing images are being created with increased refinements and data collected with various other instruments.

Ultra-Deep Field: Refined Images

In 2012, NASA released a refined version of the Ultra-Deep Field called the eXtreme Deep Field (XDF). It shows galaxies that reach back 13.2 billion years in time. We see them as they existed not long after the Big Bang which happened about 13.7 billion years ago.

The image to the right is composed with the full range of ultraviolet to near-infrared light using the Near Infrared Camera and Multi-Object Spectrometer (NICMOS) instrument on the Hubble.

Image credit: NASA and ESA

Image credit: NASA and ESA

Image credit: NASA and The Hubble Heritage Team (STScI/AURA)

Whirlpool Galaxy

The Whirlpool Galaxy can be seen with smaller telescopes. It has a near- by companion galaxy, NGC 5195, just off the upper edge of the image. The interaction between the two is spurring star formation.

Image credit: NASA and the Hubble Heritage Team (STScI/AURA)

A Backward Moving Spiral Galaxy

Galaxy NGC 4622 appears to be rotating backwards. Most spiral galaxies have arms of gas and stars that appear to trail behind as the galaxies rotate. Astronomers speculate that NGC 4622 interacted with another galaxy, and now its arms point in the opposite direction of its rotation.

Image credit: NASA/CXC/JPL- Caltech/STScI

Small Magellanic Cloud

In this image NASA's Hubble Space Telescope provided visible light data for the colors red, green, and blue. NASA's Chandra X-ray Observatory contributed X-ray data and is purple. And NASA's Spitzer Space Telescope contributed the infrared data also seen as red.

The Small Magellanic Cloud galaxy is very close to our Milky Way galaxy and is around 200,000 light-years away. It is small and orbits the Milky Way. This close galaxy gives us the chance to observe things that cannot be seen in farther galaxies.

Starburst Galaxy

NASA's Hubble Space Telescope images of starburst galaxies are used to study colors that could help find the age of these galaxies. The color in this image corresponds to the stars temperature. Young stars are blue and are hotter. While older stars are more red and tend to be cooler. The areas in the image that are bluer have more stars birthing.

NGC 3310 is located near the constellation Ursa Major, and it is about 59 million light-years from Earth. This image is compiled from observations made by the Wide Field and Planetary Camera 2 in March 1997 and September 2000.

Image credit: NASA and The Hubble Heritage Team (STScI/AURA)

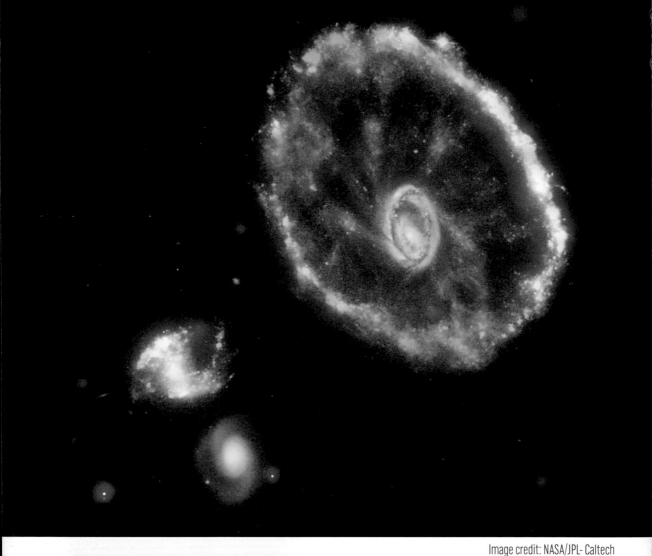

Image credit: NASA/JPL- Caltech

Cartwheel Galaxy

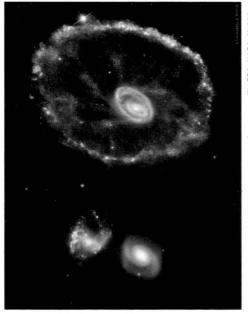

Image credit: ESA, NASA, Hubble

This wheel-shaped galaxy *(left)* is actually an interaction between two galaxies. A smaller galaxy is passing through the larger one, and the gravitational disruption is causing the ring of star formation indicated by the ring of blue in the bottom image.

The top image (of the same galaxies) is a false-color composite image made from space-based telescopes: the Chandra X-ray Observatory data is purple, the Galaxy Evolution Explorer ultraviolet view is blue, the Hubble Space Telescope visible light is green and the Spitzer Space Telescope infrared image is red. The combined data from all telescopes created a beautiful picture. It also allowed scientists to make the following deductions: a hundred million years ago, a smaller galaxy plunged through the heart of Cartwheel Galaxy; the first ripple appears as the bright blue outer ring; and, the blue color reveals that associations of stars five to twenty times as massive as our Sun are forming. The pink clumps contain binary star systems with a black hole. The yellow-orange inner ring represents the second ripple created in the collision and has very little star formation.

Star Forming Activity Spurred

This image *(below)* is of the galaxy NGC 6872. Its shape is caused by the interaction with another smaller galaxy, IC 4970, seen above it. NGC 6872 is a very large spiral galaxy. It is around five times the size of the Milky Way.

NGC 6872 has a star-forming region in its upper arm, which is indicated by the bluish area in this image. It is suggested by scientists that its interaction with galaxy IC 4970 spurred its star-forming activity.

Image Credit: ESA/NASA/STScI

Bubbles Rising from the Core

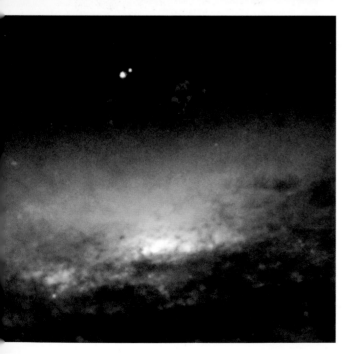

This Hubble image of Galaxy NGC 3079 was taken in 1998. Red indicates glowing gas, and blue and green indicate starlight. An enormous bubble is in the center of the saucer-shaped galaxy from which gas and the ingredients for star formation are being released in a vortex. The small image *(bottom facing page)* shows a close-up region of the galaxy's center.

Eventually, this gas will rain down on the disk and may collide with gas clouds, compress them, and form a new generation of stars.

The bubble formed, when winds from hot stars mixed with small bubbles of hot gas from supernova explosions, according to theoretical models. Radio telescope observations indicate those processes are still active. Eventually, the hot stars will die, and the bubble's energy source will fade away.

Image credit: NASA, ESA, and The Hubble Heritage Team (STScI/AURA)

Hubble-V in NGC 6822

The small, irregular host galaxy, called NGC 6822 is also home to the Hubble-V galaxy, which has a diameter of about 200 light-years. Although the galaxy is visible by ground-based telescopes, the individual young stars are not. Hubble's Wide Field and Planetary Camera 2 has the resolution and sensitivity to ultraviolet light to exhibit the nebula's cluster of hot stars. These stars are brighter than our Sun by 100,000 times.

Image Credit: NASA and The Hubble Heritage Team (STScI/AURA)

Hubble-X
Star Formation Burst

Galaxy NGC 6822 is 1,630,00 light-years away—a close neighbor to the Milky Way galaxy. It is located in the constellation Sagittarius. Hubble-X is a cloud of glowing gas and an active star-forming region inside galaxy NGC 6822. The gas clouds were discovered in 1881, but they are named after Edwin P. Hubble who made detailed photographs of NGC 6822 in 1925. This Hubble image shows the ultraviolet radiation from the massive stars within the cloud causing the gas to glow. Stellar winds are outflows of radiation and gas that will eventually disperse and end the star formation.

Messier 104

Messier 104 is referred to as the Sombrero Galaxy. It spans about 50,000 light-years across and is 28 million light-years from Earth. The cluster age ranges from 10 to 13 billion years. Astronomers have calculated that the galaxy has 2,000 globular clusters, some that can be seen in the image taken with Hubble instruments *(bottom)*. NASA's Spitzer Space Telescope added infrared information to the Hubble image *(top)*. The galaxy looks warped with the added infrared information, which is an indication that the galaxy had an interaction with another galaxy that had a gravitational impact on its symmetry.

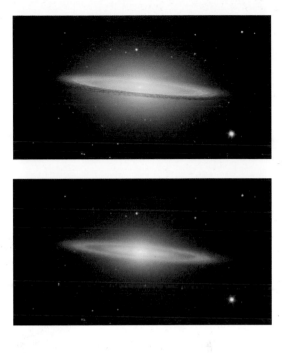

Image credits : NASA/JPL-Caltech/University of Arizona/STScI

Credit: NASA, ESA, the Hubble Heritage Team (STScI/AURA), and A. Aloisi (STScI/ESA)

A Dwarf Galaxy Puzzle:
NGC 1569

NGC 1569 *(above and right)* is a dwarf galaxy in our local neighborhood of galaxies. It has incredible star-forming activity. Its star-formation rate is 100 times more than the Milky Way galaxy. It had once

been thought that NGC 1569 was a loner or isolated galaxy. The puzzle was: how can all this star formation be happening when it's all alone, an isolated galaxy? Generally, star birthing occurs when galaxies interact with other galaxies.

Hubble supplied the answer. It was Hubble observations that discovered the galaxy to be one and a half times

farther than originally measured with ground telescopes. The new distance placed it closer to other galaxies, and it is the interaction with these other galaxies that is spurring the new stars.

NGC 1569

Is Moving Away

NGC 1569 is moving away from the Earth. It is about 11 million light-years away. At one time it was thought to be about 7 million light-years away. There are two prominent super star clusters in the galaxy where formation takes place. A dwarf irregular galaxy UGCA 92 may be interacting with NGC 1569, and it could be its companion galaxy.

Image credit: NASA/JPL/Hubble

Image credit: NASA, ESA, Dan Maoz (Tel-Aviv University, Israel, and Columbia University, USA)

Galaxy NGC 1512

Galaxy NGC 1512 is a barred spiral galaxy. The color-composite image was created from exposures taken by the Faint Object Camera (FOC), Wide Field and Planetary Camera 2 (WFPC2), and the Near Infrared Camera and Multi-Object Spectrometer (NICMOS).

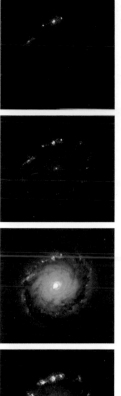
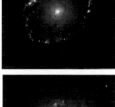

Image credit: NASA, ESA, Dan Maoz (Tel-Aviv University, Israel, and Columbia University, USA)

Wavelengths Recorded

The top two images *(left column)* show different ranges of ultraviolet wavelengths. The next two images *(middle)* are two ranges of visible wavelengths. The bottom three images are different ranges of infrared wavelengths. All seven images were combined to make the final image *(facing page)*.

Center Ring

Galaxy NGC 1512 has two rings, an outer and a center ring. This image *(bottom right)* is a detail of the center ring. The image on the following pages illustrate the distinct formation of the two rings.

Image credit: NASA, ESA, Hubble, LEGUS; Acknowledgement: Judy Schmidt

Images credit: ESA/Hubble, NASA

Merging Galaxies

This Hubble composite image *(above)* shows interacting galaxies NGC 1512 *(left)* and NGC 1510 *(right)*. These two galaxies, one much bigger than the other, will eventually be drawn closer and closer together. At some point in time they will become one galaxy. They presently have an effect on each other and are in the process of merging. Each galaxy is about 30 million light-years away from Earth. Consequently, the size seen in this image is proportional.

The image of NGC 1510 *(facing page bottom)* is a close-up from the zoomable feature on some of NASA's images. This smaller dwarf galaxy experiences a greater gravitational pull from its larger neighbor galaxy.

Image credit: NASA/ESA/CXC/SSC/STScI

Galaxy Messier 101

In these two images of Messier 101 (each formed somewhat differently), red shows Spitzer's infrared data, pointing out the heat emitted by star-forming dust. Yellow is Hubble's data displaying visible light emitted by the stars, but following the same tracks as the dust. Blue shows Chandra's recording of X-ray light. These X-ray sources are exploded stars, hot gas, and colliding materials pulled by strong gravitational forces.

Image credit: NASA/JPL- Caltech/STScI

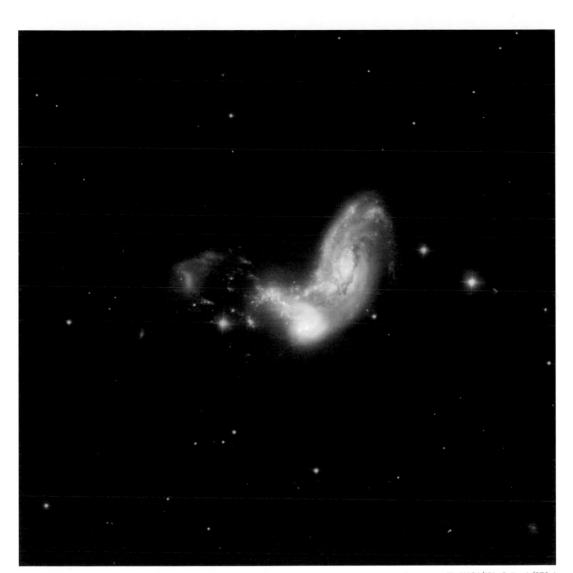

Image credit: NASA/JPL- Caltech/STScI

Merging Galaxies

II Zw 096

This image of two merging galaxies, jointly known as II Zw 096, is composed with a combination of data from NASA's Spitzer and Hubble Space Telescopes. The blue is from Hubble's far-ultraviolet and visible wavelength observations. The cyan is from Hubble's near-infrared observations. The orange is from Spitzer's infrared observations, and the red is from its mid-infrared observations.

Researchers find this galactic collision's glow is unusually far from its center. They estimate that starbursts equal to 100 solar masses are occurring each year.

A Galaxy Pair: NGC 454

NGC 454 is a pair of galaxies. One is a reddish elliptical galaxy, and the other is an irregular galaxy. Their interaction has distorted both. As a result, the clumps that are slightly blue are probably part of the galaxy that is irregular. At the time this image was made, there was no sign of star formation as a result of the collision.

Image credit: Image credit: NASA, ESA, the Hubble Heritage Team (STScI/AURA)-ESA/Hubble Collaboration, and M. Stiavelli (STScI)

Image credit: NASA, ESA, the Hubble Heritage Team (STScI/AURA)-ESA/Hubble Collaboration, and K. Noll (STScI)

Graceful Impact: Hercules Galaxy Cluster

This image is a collision between two spiral galaxies, NGC 6050 and IC 1179, which is part of the Hercules Galaxy Cluster. Together it is referred to as Arp 272. Their spiral arms seem to be gracefully entwining. A third, smaller, interacting galaxy can be seen above these two.

Collision Between Disk Galaxies

Galaxy NGC 520 is actually two disk galaxies that have collided, but whose nuclei have not yet merged. The two galaxies together measure about

Image credit: NASA, ESA, the Hubble Heritage Team (STScI/AURA)-ESA/Hubble Collaboration, and B. Whitmore (STScI)

100,000 light-years across and about 100 million light-years away from Earth. Galaxy NGC 520 is so bright it can be seen with a small ground telescope from Earth.

Image credit: NASA and the Hubble Heritage Team (STSci/AURA)

Edge View
of a Spiral Galaxy

If we could see NGC 4013 from its pole, we would see a circular pinwheel. We don't see the ends because it is larger than Hubble's field of view. There are small blue areas along the brown band *(from upper right to lower left)* that are thought to be star-forming regions. The extremely bright star in the upper left corner on the band is a nearby foreground star that is not in the galaxy at all. It is actually closer and in our Milky Way home galaxy. It just happens to be in the line of sight when viewing NGC 4013.

Image credit: NASA, ESA, and
The Hubble Heritage Team (STScI/AURA)

Cosmic Dust Bunnies

According to SpaceTelescope.org, Hubble spied cosmic dust bunnies. Just like the dust that collects in corners and under the bed, this cosmic dust shows variations and complexi-ties. We often use our earthly experiences—like dust bunnies—to describe what we see through our telescopes.

NGC 1316 is an elliptical galaxy. This image reveals dust lanes and star clusters, suggesting at one time two gas-rich galaxies had merged.

Galaxy Clusters:
Gravitational Lenses

A cosmic magnifying glass is created by the massive gravitation of huge galaxy clusters. These clusters can contain hundreds to thousands of galaxies, which are gravitationally bound to each other. Light emitted from distant galaxies behind a cluster would not typically reach us. Because of the distance, it would be too faint even for the Hubble Telescope to see. However, the gravitational effect of the cluster is strong enough to bend and magnify the light, and then, Hubble can see and record the distant galaxies. Astronomers use gravitational lensing to record and study the most far away galaxies.

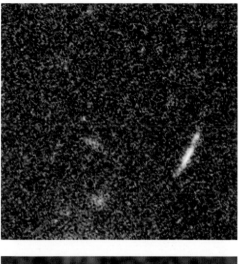

Hubble Deep Fields

One of the goals of the Hubble Deep Fields Initiative program is to study early galaxy formation using gravitational lensing.

Cluster Abell 1689

It is estimated that this galaxy cluster (*facing page*), named Abell 1689, has over a trillion stars. We are able to see remote galaxies behind it using its gravitational lens. These galaxies are arc-shaped objects that are around the cluster. The images to the right are close ups of the original images before data from different instruments are combined and before color is assigned.

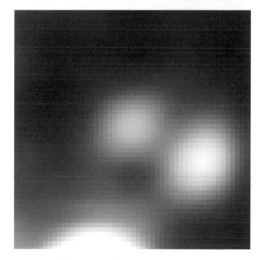

Image credit (left and right pages):
NASA/ESA/JPL-Caltech/STScI

Wide-Field Image

Compare this wide-field image of galaxy cluster Abell S1063 observed by Hubble, to the two on the facing page that do not use gravitational lensing to magnify the galaxies in this same area of the sky. A tell tail sign of gravitational lensing is the appearance of the arcs and the curved-shaped galaxies around the central cluster.

Image credit: NASA, ESA, and J. Lotz (STScI)

Ground-Based Abell S1063

This image *(top right)* was recorded with a ground-based telescope. It shows the galaxy cluster Abell S1063 and its surrounding area.

Parallel Field of Abell S1063

This image *(bottom right)* shows a part of the sky that is a parallel observation to the galaxy cluster Abell S1063. One of Hubble's cameras observed the galaxy cluster itself *(facing page)*. Another simultaneously captured this image of an adjacent area of the sky. Comparing images like this helps us to understand how similar our universe looks in different directions.

Image credit (top right): NASA, ESA, Digitized Sky Survey 2
Acknowledgement: Davide De Martin

Image credit (bottom right): NASA, ESA, and J. Lotz (STScI)

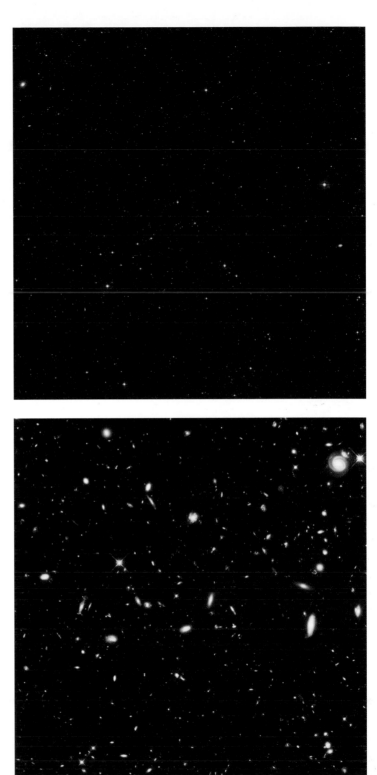

A Massive Cluster

This image of a massive galaxy cluster is called MACS J1206.2-0847. Clusters like this one work like a huge cosmic gravitational lens. Any light that enters it magnifies, distorts, and bends.

Light from distant galaxies is bent by the gravitational pull of dark matter within the massive group of galaxies that are in front of the distant ones. Astronomers, in a study called the CLASH survey (Cluster Lensing And Supernova survey with Hubble), are making maps dark matter. Dark matter cannot be seen. It can only be calculated by the gravitational effect it has on visible matter. Gravitational lenses let us see the distant galaxies beyond them that would ordinarily be too faint to observe. Galaxy clusters like MACS 1206 are ideal for extending our telescopes to faraway galaxies and for studying dark matter.

Image Credit: NASA, ESA, M. Postman (STScI), and the CLASH Team

Image Credit: NASA, ESA, M. Postman (STScI), and the CLASH Team

Frontier Fields Program

A Happy Face Cluster

The NASA/ESA Hubble Space Telescope works as part of the Frontier Fields program to capture some of the farthest reaches of the universe.

The eyes of cluster SDSS J1038+4849 are very bright galaxies. The smile and head are arcs caused by gravitational lensing. Galaxy clusters are the most immense structures in the universe, distorting and bending light.

Image credit: NASA, ESA, the Hubble Heritage Team (STScI/AURA), J. Blakeslee (NRC Herzberg Astrophysics Program, Dominion Astrophysical Observatory), and H. Ford (JHU)

Galaxy Cluster Abell 1689

Abell 1689 is a cluster of many galaxies. Visible and infrared data from Hubble's Advanced Camera for Surveys (ACS) are combined with exposure times of over 34 hours, which produces greater and more remarkable details than in earlier images. This image shows bright stars and far away spiral galaxies. Some galaxies have arcing streaks.

These streaks are signs of gravitational lensing. Abell 1689 is massive, so its gravitational forces bend or warp the space around it. This affects how light coming from objects farther behind the cluster travels through space. These streaks are the distorted forms of galaxies that lie behind the cluster. They are often too faint for us to see until they are magnified in this process of gravitational lensing—like a galactic telephoto lens.

Looking into the Past

Cluster MACS J0647+7015's gravitational lens makes visible one of the farthest know galaxies, MACS0647-JD. This galaxy would have been invisible without this galactic zoom lens. It is a young and small galaxy—smaller than our Milky Way.

The universe is about 13.7 billion years old. This galaxy, as we see it here, is what it looked like when the universe was only three percent of that age. The light took that long to reach the galaxy cluster lens and then to reach the Hubble's lens.

Image credit: NASA/ESA/STScI/CLASH

Image credit: NASA, ESA, and T. Johnson (University of Michigan)

Cluster
SDSS J1110+6459

This cluster known as SDSS J1110+6459 is 6 billion light-years from Earth. It contains hundreds of galaxies. The blue arc is made up of three separate images of an even farther galaxy called SGAS J111020.0+645950.8. This far away galaxy has been magnified, although also distorted, by a gravitational lens provided by the massive galaxy cluster SDSS J1110+6459.

Image credit: NASA, ESA, D. Harvey (École Polytechnique Fédérale de Lausanne, Switzerland) and R. Massey (Durham University, UK)

Dark Matter Map Galaxy Cluster

A map of the dark matter found within the cluster is shown in blue. This cluster was in a study of 72 galaxy cluster collisions. It found dark matter does not interact with other dark matter as much as was once thought.

Image credit: NASA/ESA/STScI

Abell 2744 Cluster Lens

For this image, Hubble once again reached into the deepest of space with the help of massive galaxy cluster Abell 2744. Abell 2744's lens warped space, brightening and magnifying images of almost 3,000 distant background galaxies. The light took over 12 billion years to reach us. In this image, we are seeing the galaxies as they existed 12 billion years ago.

Image credit: NASA, ESA, M. Montes (IAC), and J. Lotz, M. Mountain, A. Koekemoer, and the HFF Team (STScI)

Intracluster Space Light

The glimmer of blue light, not coming from stars or galaxies, is coming from dead galaxies that have been torn apart because of the cluster's gravitational forces. The space between galaxies is called intracluster space.

Multiple Galaxy Clusters

Astronomers study the gravitational effects that the cluster content has on the light—distant light from remote objects that lie farther beyond them.

They map data collected by Hubble observations about the mass inside the clusters.

Image credit: ESA/Hubble, NASA, HST Frontier Fields
Acknowledgement: Mathilde Jauzac (Durham University, UK and Astrophysics & Cosmology Research Unit, South Africa) and Jean-Paul Kneib (École Polytechnique Fédérale de Lausanne, Switzerland)

Mapping Mass

The blue haze on the image above was created from the mass map *(left)*. In reconstructing where the mass is located, the general area of where the invisible dark matter is can be calculated. This calculation is then applied to the image and communicates the dark matter's presence.

The Future of Space Telescopes

The chief function of a space telescope isn't to take spectacular photos. The purpose of space telescopes is to study and learn about the universe. However, I like to think creating amazing images is one of their main roles. It puts our discoveries in context and makes the things in our universe tangible.

Space telescopes, like the forthcoming James Webb Space Telescope, will have even greater capabilities to reach farther in to space with even greater resolution. No doubt, it will outperform and out-discover anything that came before it, including the Hubble Telescope. Much of its collected data will not be in the visible light range, but the images are certain to be as dazzling as those of its predecessor, Hubble.

We have much to look forward to in the coming years.

Index

A

aurorae, 37

B

big bang theory, 63, 85
black holes, 5, 19, 44, 56,
57, 58, 59, 65, 68, 70,
71, 82, 91
brightness, 14-15, 32,
44, 53, 59

C

Chandra X-Ray Observa-
tory, 16, 17, 45, 59, 70,
71, 88, 91
clouds, planetary, 21,
33, 36, 39,
clouds, galaxic, 46, 48, 49,
68, 73, 76, 88, 93, 94
color, 16-17, 20, 31, 32,
33, 45, 47, 52, 58, 88,
89, 91, 111
color, true and false, 19,
45, 52, 91
color composite, 11, 16,
19, 30, 32, 91, 98
color mapping, 31
Comet Shoemaker-
Levy 9, 29, 30
comets, 19, 29, 30,
31, 42, 43
comet impacts, 29, 30
comets, sungrazing, 43

D

dark energy, 60-63
dark matter, 60-63, 82,
114, 119, 123
distance, 11, 14, 18, 21,
61, 67, 97, 111
dust shells, 51

E

European Southern Ob-
servatory (ESO), 59, 68

G

galactic telephoto lens,
116
galaxies, merging/
interacting, 82, 86,
90, 91, 96, 97, 100,
103, 105, 119
galaxy evolution, 12,
57, 83, 91
Galileo Galilei, 7, 14
gas, 12, 15, 33, 43,
gas, interstellar, 47, 51,
52, 54, 47, 59, 70, 72,
73, 74, 75, 81, 82, 82,
87, 92, 93, 94,
102, 109, 118
globules, 48
gravitational lensing, 63,
111, 112, 114, 116
grayscale images, 31,
16, 43, 47
Great Red Spot, 29

H

Hubble, Edwin, 7, 62,
65, 82, 94
Hubble Space Telescope,
10-19, 20-22
Hubble Telescope
Instruments
Advanced Camera for
Surveys (ACS), 9, 12,
13, 116, 117
Cosmic Origins
Spectrograph
(COS), 12, 13
Fine Guidance Sensor
(FGS), 12, 13
Near Infrared
Camera and
Multi-Object
Spectrometer
(NICMOS), 12,
13, 85, 98
Space Telescope Imaging
Spectrograph (STIS;
non-operative), 12, 13
Wide Field Camera 3
(WFC3), 12, 13, 29,
32, 38, 46, 48, 75,
89, 93, 98

I

infrared, 12, 13, 14, 16,
19, 39, 45, 52, 68, 70,
71, 72, 76, 79, 85,
88, 91, 95, 98, 99,
102, 103, 116

instruments, 5, 7, 8, 9, 12, 13, 16, 18, 19, 34, 61, 84, 95

J

James Webb Space Telescope, 9, 125

L

light and radiation, 12, 33, 44, 47, 52, 55, 94
light years, 5, 18-19
luminosity, 14

M

mapping, mass and matter, 123
Milky Way, 54–81, 82, 88, 91, 94, 96, 108, 117
moons,
    Charon, 18, 40, 41
    Europa, 28, 31
    Hydra, 40, 41
    Nix, 40, 41
    Earth's Moon, 10-11
    Europa, 28, 31

N

nebulas
    diffuse, 75,
    galactic, 93
    planetary, 50, 51, 52, 54, 65, 75
    protoplanetary, 49, 54
    ring, 47

P

pixels, 19, 21
planets
    Earth, 22
    Jupiter, 28-31
    Mars, 24-27
    Mercury, 20
    Neptune, 38-39
    Saturn, 32-35
    Uranus, 36-37
    Venus, 21
poles, planetary, 24, 37
pulsars, 55

R

resolution of images, 11, 19, 21, 74, 75, 125
rings, planetary, 32, 33, 34, 35, 36
rings, galaxy, 99

S

satellites, 22, 36, 40, 31
service missions, 7, 8
stars, 12, 14, 15, 18, 40, 43, 44, 46, 48, 53, 57, 63, 66, 68, 69, 70, 73, 76, 77, 79, 81, 82, 87, 89, 91, 93, 94, 97, 102, 111, 116, 121
solar systems, 5, 18, 20, 27, 40, 42, 65, 80
solar wind, 37
Spitzer Space Telescope, 16, 44, 45, 52, 70, 88, 91, 95, 102, 103
star formation, 46, 63, 73, 75, 80, 86, 90, 92, 93, 94, 96, 108, 111

stellar magnitude, 14
supernovas, 40,, 44, 45, 46, 55, 63, 67, 72, 93, 114
supernova remnants, 45, 55, 67

T

telescopes, 5, 7, 11, 14, 18, 44, 45, 55, 61, 70, 78, 86, 91, 93, 97, 103, 109, 114, 124, 125

U

ultraviolet light, 11, 12, 16, 19, 21, 29, 31, 49, 52, 91, 93, 94

W

wavelength, 7, 12, 14, 15, 16, 39, 61, 70, 99, 103
weather, 22, 25, 38, 39
white dwarfs, 50, 52, 77
wide-field camera, 12, 29, 32, 38, 46, 48, 75, 89, 93, 98
wind, planetary, 29, 39,
winds, 71, 93, 94

X

x-ray, 16, 45, 57, 59, 70, 71, 88, 91, 102

Z

zoom, galactic, 116

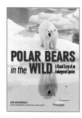

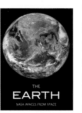

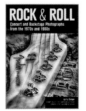
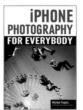